High School Mathematics

Reimagined Revitalized and Relevant

A Publication of

1906 ASSOCIATION DRIVE | RESTON, VA 20191-1502

Director of Publishing and Creative Services: Scott Rodgerson
Production Manager: Sandy Jones
Editorial Production Manager: Stephanie Levy
Developmental Editor: Myrna Jacobs
Cover and Interior Design: Kirsten Ankers
Layout: KnowledgeWorks Global, Ltd.

The National Council of Teachers of Mathematics advocates for high-quality mathematics teaching and learning for each and every student.

High School Mathematics Reimagined, Revitalized, and Relevant is an official position of the National Council of Teachers of Mathematics, as approved by the NCTM Board of Directors, April 2024.

Recommended citation:
National Council of Teachers of Mathematics. (2024). *High School Mathematics Reimagined, Revitalized, and Relevant.*

ISBN: 978-1-68054-148-9
ISBN (ebook): 978-1-68054-149-6
Library of Congress Control Number: 2024943465

Printed in the United States of America

Contents

High School Mathematics Reimagined, Revitalized, and Relevant Writing Team

Kristi Martin, Chair
Tumwater School District
Tumwater, Washington

Kris Cunningham
Phoenix Union High School District
Phoenix, Arizona

Christine Franklin
American Statistical Society
University of Georgia
Athens, Georgia

Dewey Gottlieb
Hawaii State Department of Education
Honolulu, Hawaii

Kandi Hunter
Region 10 Education Service Center
Dallas, Texas

Karen Hyers
Tartan High School
Oakdale, Minnesota

Brian Lawler
Kennesaw State University
Kennesaw, Georgia

Jenny Novak
Howard County Public School System
Ellicott City, Maryland

Judith Reed Quander
University of Houston-Downtown
Houston, Texas

Lya R. Whiting Snell
Georgia Department of Education
Atlanta, Georgia

Michael Steele
Ball State University
Muncie, Indiana

David Barnes, Staff Liaison
National Council of Teachers of Mathematics
Reston, Virginia

Contributing Writers

Kevin Dykema
Mattawan Middle School
Mattawan, Michigan

Latrenda Knighten
East Baton Rouge Parish Schools
Baton Rouge, Louisiana

Steering Committee

NCTM Board of Directors

ASSM Board of Directors

NCSM Board of Directors

Preface

Does it seem like change is happening more rapidly? It does, and it is. While we have been working and promoting statistical problem-solving for years with our colleagues at the American Statistical Association, data science and big data have become areas of need and topics of intense interest. When we started developing this publication, artificial intelligence was not a reality for most people. These technological advancements highlight the importance of reimagining and revitalizing high school mathematics to make it relevant and valuable for all students, leading to future generations who comfortably and purposefully embrace mathematics.

Progress in mathematics education is happening because of the work of teachers in the classrooms, who are often supported by programs, policies, and changes championed by school district leaders and state-level efforts. Still, the challenges persist, the progress is uneven, and the gains are often inequitable.

This work and the progress we have made in understanding student capabilities, seeing what student engagement in learning and doing math looks like, and preparing students for the world are based on decades of input from math teachers and leaders. The vision and guidance set forth in An Agenda for Action (1980) are still relevant today. The Principles, Process Standards, and Mathematical Practices of the 1989 and 2000 National Council of Teachers of Mathematics (NCTM) Standards, coupled with Focal Points (2006) and corresponding Focus in High School Mathematics (2009), have shaped mathematics education in the United States and Canada over the past four decades, influencing state and provincial standards, curriculum materials, and professional development.

Building on this foundational work, NCTM, NCSM: Leadership in Mathematics Education, and the Association of State Supervisors of Mathematics came together to create a writing team to continue to move our thinking forward. Mathematics teachers and leaders collaborated to craft a thoughtful road map for advancing high school math education, using the student perspective as a guide. The work presented here is not the final step but only the next step in our efforts to improve mathematics learning for all high school students and positively impact the teaching experiences of high school mathematics teachers.

We recognize all those on the Writing Team for their contributions. We acknowledge and appreciate the leadership of Judith Quander, who led this effort during the initial phases of this project, and Kristi Martin, who was the Writing Team chair through the development, revision, and finishing stages of this effort.

This collaborative effort provides additional guidance to the mathematics education community on the types of experiences that students should have during their high school years. Our work must continue to help students see the relevance of what they are learning. This can only be done by reimagining content and course structures, which will ultimately help students realize revitalized experiences and provide them with opportunities to develop mathematical and statistical processes.

This work is vital for everyone involved in mathematics education and education in general. We owe this effort to our students, and we are obligated as mathematics educators to continue this collaboration. We look forward to seeing the impact of this work and will continue to provide support to enact the vision set forth here.

Kevin Dykema
President, 2022–2024
National Council of Teachers of Mathematics

Lisa Ashe
President, 2022–2024
Association of State Supervisors of Mathematics

Katey Arrington
President, 2023–2025
NCSM: Leadership in Mathematics Education

Acknowledgments

The *High School Mathematics Reimagined, Revitalized, and Relevant* Writing Team sought and received significant and substantial comments for early outlines and subsequent drafts of the manuscript. The Writing Team is especially indebted to the reviewers, named and anonymous, for their feedback and critiques. These reviewers included teachers, school leaders, mathematicians, researchers, curriculum developers, policymakers, and mathematics educators.

This book was significantly influenced by the views and perspectives shared through the review process. NCTM expresses sincere thanks to all reviewers who offered thoughtful responses. Although their contributions do not constitute an endorsement of the final version, the perspectives and insights they provided have contributed to a better publication.

The NCTM Board of Directors and the *High School Mathematics Reimagined* Writing Team are grateful to all who reviewed and shared their expertise, and we also extend thanks for the timely, thorough, and thoughtful editorial work of Myrna Jacobs.

Reviewers

Robert Berry
Shakiyya Bland
Gail Burrill
Kyndall Brown
Natevidad Casas
Ted Coe
Arlene Crum
David Dai
Zandra de Araujo
Dave Ebert
Alison Espinosa
Tim Flatley
Mark Freed
Joanie Funderburk
Karen Graham
Scott Hendrickson
Maria Hernandez
Joleigh Honey
George Hurlburt
Paul Kehle
Christine Koerner
Dave Kung
Yvonne Lai
Matt Larson
Abi Leaf
Hollylynne Lee
Steve Leinwand
Travis Lemon
Cathy Martin
W. Gary Martin
Michael Martirano
Tina Mazzacane
Rebecka Peterson
Mary Pittman
Catherine Roberts
Pam Seda
Cathy Seeley
Ryann Shelton
John Staley
Charlie Steinhorn
Marilyn Strutchens
Erin Sylves
Zalman Usiskin
Crystal M. Watson
Jeff Weld

NCTM extends thanks to our anonymous reviewers.

CHAPTER 1

Centering the Student—Reimagined, Revitalized, and Relevant Mathematics Learning

"As much as I want my students to learn about derivatives and integrals, what I really want them to take away is that they're the authors of their own story and that they are inspired to make sure that the light always has the final word in their chapters."

—Rebekah Peterson
2023 National Teacher of the Year

What are the enduring understandings and beliefs students take from their experiences in kindergarten through Grade 12 (K–12) mathematics classrooms?

When students leave high school, how many have their mathematics stories end in light? Believe in themselves? Believe in their abilities? Believe in their mathematical future? Or do students leave believing they are not proficient at mathematics? That mathematics is not useful? That learning mathematics was a test of endurance and perseverance, a filter with no other significant value?

Current high school mathematics teachers, mathematics leaders, and education policymakers did not create the system that is our high school mathematics education. None of us. We inherited over 100 years of history, instantiating and reinforcing the policies, practices, processes, and structures of today. For nearly all of us, these policies, practices, and structures result in the majority of students leaving high school without the mathematical understanding and skills deemed necessary (National Research Council [NRC], 2001; RAND Corporation, 2002) and believing they are not good at mathematics, are not capable of learning mathematics, and there is little or no relevance to mathematics or mathematical thinking they have learned. Even those who are deemed successful in high school mathematics often have the goal of never having to take another mathematics class.

In this publication, our aims are clear: to increase our commitment and effort to create a reimagined high school mathematics in the service of students, building on the progress we have made since publishing *Catalyzing Change in High School Mathematics: Initiating Critical Conversations* (National Council of Teachers of Mathematics [NCTM], 2018). This seminal publication describes mathematics systems and structures that help students develop the questioning, skills, and practices to make sense of the world around them. These systems and structures should focus on mathematics that students find worthy of their time, energy, and effort. Opportunities must exist for each and every student to leave high school with an increased belief in their mathematical ability and potential.

PAUSE AND REFLECT: Taking Stock of Our Educational Environment

- What mathematics courses and content do your students want?
- What proportion of students graduate with 2, 3, or 4 years of high school mathematics?
- For students who do not finish high school, what are their mathematical learning experiences?
- What do students choose to do after high school graduation, and what mathematics do they need?
- What mathematics courses are currently offered for all students? Some students?
- What are the student characteristics of each course? What data are being used to determine placement in those courses?
- What mathematics should your school emphasize for high-demand jobs and fields of study? Who are the experts in your community that can help support this work?

Transforming Mathematics From the Student Perspective

When we think about mathematics standards, progressions, problems, and instruction, it is easiest to focus on the content and minimize or disregard the role of students. The structure and connections of mathematics content make it possible, from an academic perspective, to consider how to sequence the content, algorithms, techniques, and skills through a progression of problems and exercises to build from point A to point B. If we are truthful, in too many cases, given the many pressures of the system, we find ourselves in a situation where we feel professionally obligated to just "cover the material."

Creating Lasting Value in Mathematics

High school students are pragmatic when it comes to studying—Why is this relevant? When is this going to be useful to me? What is it you want me to do? Isn't there an app for this?

When students perceive most of their time spent on meaningless work constructed out of a system of arbitrary rules, it is important to acknowledge that high school mathematics can be a demeaning experience for far too many students. It is equally important to acknowledge that only 40% of students (ACT, 2019) are estimated to have been successful in high school mathematics and sufficiently prepared for acceptance into a desired college or university. However, a much smaller percentage of successful students are interested, prepared, or motivated to pursue degrees requiring advanced mathematics or mathematics of any kind, and there is a decline in those interested in STEM (science, technology, engineering, and mathematics) careers.

Rather than accepting the persistent reality that many students become increasingly alienated by mathematics as they progress from course to course, we must reimagine the high school mathematics experience so that it provides all students with opportunities to "embrace the wonder, power, and responsibility of mathematics by nourishing [their] affection for it" (Su & Jackson, 2020, p. 8). How can high school mathematics experiences be designed to develop and support students in asking mathematical questions about the world around them and critically thinking about ways to find answers to those questions? Could students' mathematical learning experiences support their curiosity and creativity? Could students' mathematical learning experiences be designed so that they do not induce anxiety or feelings of inferiority?

One goal should be for each and every student to leave their K–12 education feeling positive or even neutral about learning and doing mathematics and receptive to continue pursuing mathematics or a line of study or work that includes and uses mathematics. Students should see themselves as mathematical learners and doers, which includes engaging in questioning and sense making, coupled with problem exploration and solving skills that can be applied to use statistics, quantitative literacy, and mathematical modeling to answer their questions and describe areas of interest or concern. Students who engage in learning that is grounded in experiences use mathematics to ask questions and explore varied approaches to consider, analyze, and potentially solve problems, recognizing that in some real-world questions, it is difficult to tell if they have solved the problem. This type of mathematics learning, sense making, and application contributes to a positive self-efficacy toward mathematics.

Content Relevancy

The content we expect students to learn and the means through which we engage them must reflect the needs and goals we have for students in today's world. Organizing the mathematics content and creating learning experiences to cultivate and serve the goals students have can foundationally change mathematics learning.

The impetus to align high school mathematics with the needs of students for living in today's world must be more than to ensure that they enroll in college. Two thirds of the class of 2020 high school graduates enrolled in a college or university (U.S. Bureau of Labor Statistics, 2021). By default, a mathematics learning experience solely focused on college admissions failed one third of the students. High school graduates also pursue trade certificates and apprenticeship programs or enter the workforce. Our high school mathematics programs should serve these students as well and not alienate them.

Potential trajectories after high school compel us to include quantitative, statistical, and probabilistic thinking as an integral part of the mathematics curriculum, along with the ability to analyze, critique, and develop "the visual exploration of real-world data [which] also open new possibilities for a deeper understanding of society" (Andre & Zsolt, 2019). The proliferation of technologies makes computational answer-getting nearly obsolete and puts significantly more value on problem development, questioning, analysis, interpretation, and communication. All of these actions are grounded in advanced mathematical thinking with technology (Dick & Hollebrands, 2011; Sacristán, 2021).

Mathematics Usefulness

We must revitalize high school mathematics to ensure that all students are prepared with mathematical knowledge, skills, and mindsets so that they are ready to take advantage of opportunities and navigate the challenges of living in the world today. Students' mathematics experiences should cultivate their talents, character, and aspirations, allowing them to experience joy and develop important habits of mind for seeing, questioning, and applying mathematics in their lives and in the world around them.

Focusing on student-centered desired outcomes is critical to the work of designing high school mathematics learning experiences that serve every student. As part of the enduring impact of the high school mathematics experience, all students should be able to:

- be prepared to use mathematics to help navigate the challenges of living in today's world;
- use mathematical reasoning, communication, and social interaction skills to be critical consumers of information and to defend or refute a claim;
- see mathematics as a valuable tool to question and make sense of the world, inform decisions, and support strategies and solutions in their personal lives as well as in their jobs and careers;
- be equipped with the knowledge, skills, and mindsets to be able to take advantage of opportunities including, but not limited to, postsecondary education programs; and
- see themselves as capable of learning, questioning with, and using mathematics and able to experience joy in these endeavors.

Engaging Pedagogy

The pedagogy of high school mathematics has remained static, with teacher-directed whole-group instruction focused on procedures dominating our high school landscape (NCTM, 2018; Banilower et al., 2018). These practices stand in stark contrast to research-based effective mathematics teaching supported by cognitive science and mathematics education research (Boston et al., 2017; NCTM, 2014, 2018).

Students' interest in mathematics and their beliefs about its utility for their future continue to decline from middle school to high school (Irwin, et al., 2023). Further, the public perception of the usefulness and importance of high school mathematics, as well as who is and is not positioned to be successful in high school mathematics, is a significant impediment to progress that must simultaneously be addressed (Herbel-Eisenmann et al., 2016; Shah, 2017; Wagner, 2019).

Much of high school mathematics teaching can be characterized by "I do. We do. You do." or students watching a video and then mimicking what they saw. Engaging mathematics is about students actively working to make sense of what they are doing. At the root of sense making, a person explains their own thinking and critiques the thinking of others. This requires that students talk and listen to other students and analyze what they are saying. For many students, this rarely happens in their high school mathematics courses.

In addition, it is common to hear a mathematics teacher say, "We just need to cover this material." Coverage is too often the process of presenting mathematics content, procedures, or algorithms to fulfill a requirement or standard with little expectation of student learning, understanding, or skill development. Covering also communicates to the student that this experience is about compliance and not learning.

Student Curiosity

Mathematics learning has the potential to cultivate students' natural curiosity to ask questions and build knowledge. Mathematics and data permeate the world around us. A key to increasing levels of student engagement is connecting mathematical ideas to real-life contexts. The existing landscape of mathematics learning often lacks meaningful connections to the real world. In the absence of these connections, the widespread belief that mathematics is detached from practicality and irrelevant will continue to permeate society.

The "stop-thinking questions" (Liljedahl, 2021) students commonly pose, such as the often-encountered "Is this right?," underscore a prevalent mindset within the mathematics classroom that places a disproportionate emphasis on solutions as the most valuable aspect of mathematical exploration. Real-life problems rarely conform to the tidy one strategy, one solution framework often presented in traditional classrooms. The common single strategy, single-answer approach, when used exclusively, ignores the complexities of real-world decision-making and reinforces the belief that mathematics is irrelevant.

With increased access to global information and events, students have become more attuned to social issues and global challenges. This heightened awareness provides educators with the opportunity to harness these real-world concerns and utilize them as catalysts for engaging students in developing their understanding and application of mathematics. Global challenges require the sort of problem-solving practices and competencies that have long been promoted by the mathematics education community (Boston et al., 2017; CCSSO 2010; NCTM, 2000, 2009).

Factors Influencing High School Mathematics

Technology in High School Mathematics

Technology continues to gain power and influence in nearly every phase of our lives, especially with the increasing development and use of artificial intelligence (AI). How this continuing advancement impacts education and mathematics teaching and learning provides not only new opportunities but also new challenges.

Mathematics teaching and learning continues to grapple with and embrace the question of how to incorporate technological tools to learn mathematics and apply mathematics from calculators and graphing calculators to computer algebra system (CAS) and dynamic geometry to AI. Central to these challenges and opportunities are how to learn with, think with, question with, and communicate with mathematics better using these tools. How does technology increase our student's ability and facility to use mathematics to visualize, analyze, and solve real and relevant problems? It also raises the question of where we need more emphasis and less or no emphasis, and what those implications are.

Relevance of Statistical Thinking and Data Science

Statistical thinking, reasoning, and related skills are significant aspects of what many describe when they discuss data science. A facility to understand, interpret, and question statistics and data science contexts and statements is becoming increasingly important for students. Data are the currency of and driver for communicating, decision-making, and being involved with social and commercial trends. Our goal is to prepare students to be active and informed members of society, and this requires that they leave high school with the ability to interpret and analyze statistical statements and arguments and question and reason statistically. Data science expands this domain to include elements of quantitative literacy, computational thinking, coding, and ethics (NCTM, 2024).

The increased application of big data and technologies to mine data makes understanding the implications and ethical use of large data more necessary. Whether the data set is small or large, primary or secondary, numbers or photos, or a simple or complex analysis, it is vital that students understand and can apply the statistical problem-solving process to interrogate and investigate questions, statements, and applications.

High-Stakes Assessment

For many students, what persists from high school mathematics is not the content, skills, and strategies they acquired; for many, those are long forgotten, but the opportunities or denial of opportunities influenced by a mathematics assessment lingers.

The emphasis on standardized test scores has placed pressure on teachers to ensure that students memorize procedures rather than foster their ability to engage with and reason about fundamental mathematical concepts. This is in direct conflict with an approach that centers the student and their needs and desires for learning mathematics in high school (Amrein & Berliner, 2003).

The nature of high-stakes assessments is to filter out students who are not able to graduate. Or be considered for acceptance into a university. Or be considered for a specific major. Unfortunately, the goal of these high-stakes assessments is not to impart enduring skills but to create a hurdle to be crossed and then forgotten or worse, become a lasting failure. This educational system challenge includes high school counselors and extends to college and university admission professionals and both formal policies and applied practices, as well as the assessment community. We must design high school mathematics to empower students with enduring skills to think critically, comprehend the world around them, and broaden their opportunities to take into account the larger impacts of high-stakes assessments.

Choosing Curriculum

Curriculum and pedagogy are connected. Ultimately, we want the mathematics learning experienced by students to be relevant, useful, and engaging, and curriculum can contribute to this work. An integrated curriculum can provide additional opportunities to make connections within and across mathematical content areas and combine mathematics content to support the questioning and understanding of real contexts (Grouws et al., 2013).

No matter the structure, students must be actively engaged in meaningful mathematics learning. Both traditional and integrated structures have the potential to center the student (the desired outcomes we have for our students), define the enduring learning we want students to walk away with after 4 years of high school mathematics, and then identify and organize content into course pathways accordingly.

Moving Ahead

The work to revitalize high school mathematics will not be easy. History has shown that.

The role that mathematics plays in increasing students' opportunities or restricting them will only grow. We are stronger and more productive as we increase our understanding of mathematics, our ability to ask and answer mathematical and statistical questions, and our use of mathematics to make sense of and attack the problems in the world around us.

Although this work requires many individuals, we need to make systemic change. To better meet the needs of all high school students, we need change at the school, district, state, and national levels. We need to engage every high school student in a mathematics learning experience that is based on sense making, problem-solving, and using practices and skills to question and generate solutions to meaningful contexts.

This task requires every one of us. What follows is not a recipe but a framework to inspire a fresh perspective on our actions, the value of content for students, and the practicality of mathematics. Our goal is to foster generations of students who leave with the conviction that they can understand, learn, and apply mathematics to explore their own interests based on the following core actions:

1. **Relevance becomes a defining characteristic of mathematics classrooms and learning** through mathematical and statistical modeling and the use of contextual and interesting tasks.
2. **Reimagine the content by organizing around Crosscutting Concepts and creating interest-driven pathways** to deepen their understanding of mathematics and statistics, highlight connections between the mathematical and statistical concepts, and allow students to recognize the overall utility of the mathematics they are learning.
3. **Revitalize the student experience by using mathematical and statistical processes** to ensure student engagement, active sense making of mathematics and statistics, and use of mathematics and statistics to question, understand, or critique the world around them.

PAUSE AND REFLECT: Considering the Student Experience

- Which students are currently benefiting from your current system and structures? Which students are not?
- Do your students currently leave high school recognizing the relevance and usefulness of the mathematics they learned?
- Do the skills that your students are learning prepare them for success in their post-high school endeavors?
- What enduring mathematics skills and practices do you want every child to leave high school with?

CHAPTER 2

Relevance: Engaging Students in Mathematical and Statistical Modeling

A key to increased levels of understanding and engagement is using mathematics to ask and answer real questions in real contexts. Engaging in mathematics learning through real-life contexts better prepares students to ask and answer questions mathematically and provides mathematics learning experiences that connect to subjects, contexts, and possible career fields. Students should experience using mathematics as a tool to explain, analyze, investigate, and model situations that exist in life. They should have consistent opportunities to solve meaningful, real-life problems and know when to use mathematics as a vehicle to make sense of authentic situations and as a tool to answer *their* questions. Put simply:

> Modeling is the engine that drives mathematical and statistical inquiry and learning in high school.

Mathematical modeling encompasses ideas such as problem-solving and real-world applications, which are critical features of a mathematics learning program that develops students' capacities to use mathematical tools in meaningful ways. Although some individual definitions may differ, mathematical and statistical modeling goes beyond contextual problem-solving and applications to include a cycle of activities that involve determining which components of a situation to mathematize and developing explanations of how creating models, interpreting results, and revising models to attend to different factors ultimately improve the efficacy of the model (Abassian et al., 2020; Cirillo & Pelesko, 2022; Hirsch & McDuffie, 2016).

Providing both content and learning experiences that are authentic and rich provides an opportunity for students to see the relevance, power, and beauty of interconnected mathematical ideas. Ideally, learning experiences should provide opportunities for students to discover the joy and wonder of mathematics and to develop patience and persistence when they solve problems. Mathematics, statistics, and data science impact life all around us. Mathematical modeling, statistical reasoning, and data literacy are important topics for all high school mathematics to make connections with all real-life phenomena. These connections should be made within and between courses.

PAUSE AND REFLECT: Taking Stock of Your Students' Interests

- Are students actively making sense of the mathematics and statistics they are learning?
- How often and when do students see the usefulness of the mathematics and statistics they are currently learning?
- As students are learning mathematics, are they making connections within mathematics?
- Are students applying mathematics and statistics to solve real-world problems?

Focusing on Mathematical and Statistical Modeling

To support high school graduates who are adequately prepared for a technologically changing society, high school mathematics must build students' abilities to define and develop questions, think critically, reason with mathematics, and adapt to novel situations. To achieve this goal, modeling must be a foundational activity in mathematics and statistics high school coursework. The need to make high school mathematics modeling centric, relevant, and grounded in real-world problem-solving is not new. It is, however, even more important than ever.

Previous calls for reform in high school mathematics have not meaningfully shifted the canon of practice at the national level. Moreover, the ability to address significant global challenges such as public health (e.g., Ndaïrou et al., 2020), climate change (e.g., Hartzell et al., 2023; Kim et al., 2019; Sun et al., 2022), global economics (e.g., Sana, 2022), and AI (e.g., Ferreira, 2006) are grounded in mathematical and statistical modeling. As noted in the CCSSO's High School Modeling domain (2010), "Modeling links classroom mathematics and statistics to everyday life, work, and decision-making. Modeling is the process of choosing and using appropriate mathematics and statistics to analyze empirical situations, to understand them better, and to improve decisions."

The mathematical foundations laid in middle-grades mathematics and statistics position students well to engage in mathematical and statistical modeling throughout their high school coursework. Students enter high school with an understanding of the nature of linear and exponential functions, opportunities to summarize center and spread in data sets, and experience with geometric shapes and solids and related properties. These tools provide a foundation to begin modeling meaningful real-world situations and data sets. Moreover, modeling complex situations motivates the need for new families of functions (such as quadratic, rational, and trigonometric), new statistical tools (such as residuals, inferential testing, and exponential and logarithmic regression), and the deepening of geometry and measurement properties and reasoning structures.

When high school teachers position modeling as a focus of mathematics and statistics learning, students have opportunities to critically analyze situations, determine which features are important, develop mathematical or statistical models, test the models' efficacies in making predictions, and revise and iterate the models. The mathematical and statistical modeling cycles become visible components of every high school course. Mathematical and statistical modeling applies to each strand of mathematics, including algebra and functions, geometry and measurement, and statistics and data science.

This process should not be an end unto itself but a means for students to mathematize situations to make inferences, estimates, predictions, and conclusions that inform action. Modeling is not a focus or unit that fits solely into one course or mathematical strand; rather, it is a process that can and should be used to engage students in using and developing their understandings and skills and motivate their need for new content to answer new questions of interest.

Modeling Cycles

The modeling cycle can be seen through either a mathematical lens, with a focus on algebra, functions, geometry, and measurement, or a statistical lens, attending to statistics and probability situations and data science contexts. Figures 2.1 and 2.2 portray the modeling cycle through a mathematical lens and a statistical lens. The early stages of the modeling cycle involve identifying the key quantities in a situation. In the context of mathematics, this involves identifying the problem to be solved, defining key

Figure 2.1
The Modeling Cycle (Mathematical Lens)

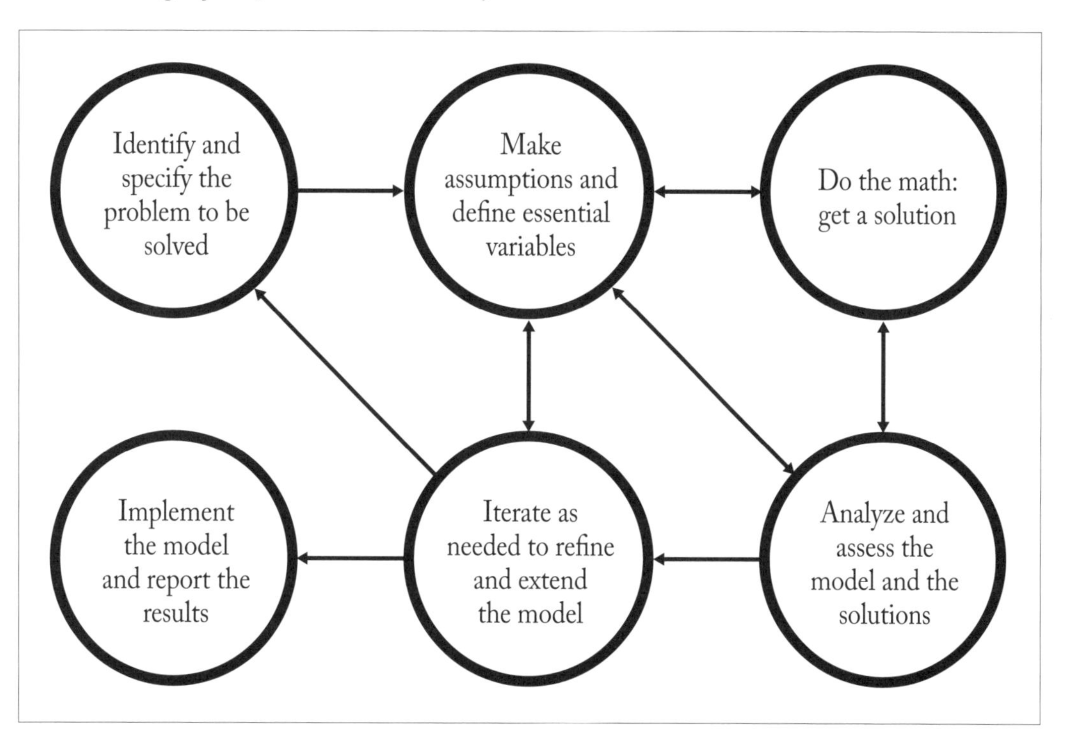

Source. Garfunkel, S., & Montgomery, M. (Eds.). (2019). *GAIMME: Guidelines for Assessment and Instruction in Mathematical Modeling Education* (2nd ed.). Consortium for Mathematics and Its Applications (COMAP) and Society for Industrial and Applied Mathematics (SIAM).

Figure 2.2
The Modeling Cycle (Statistical Lens)

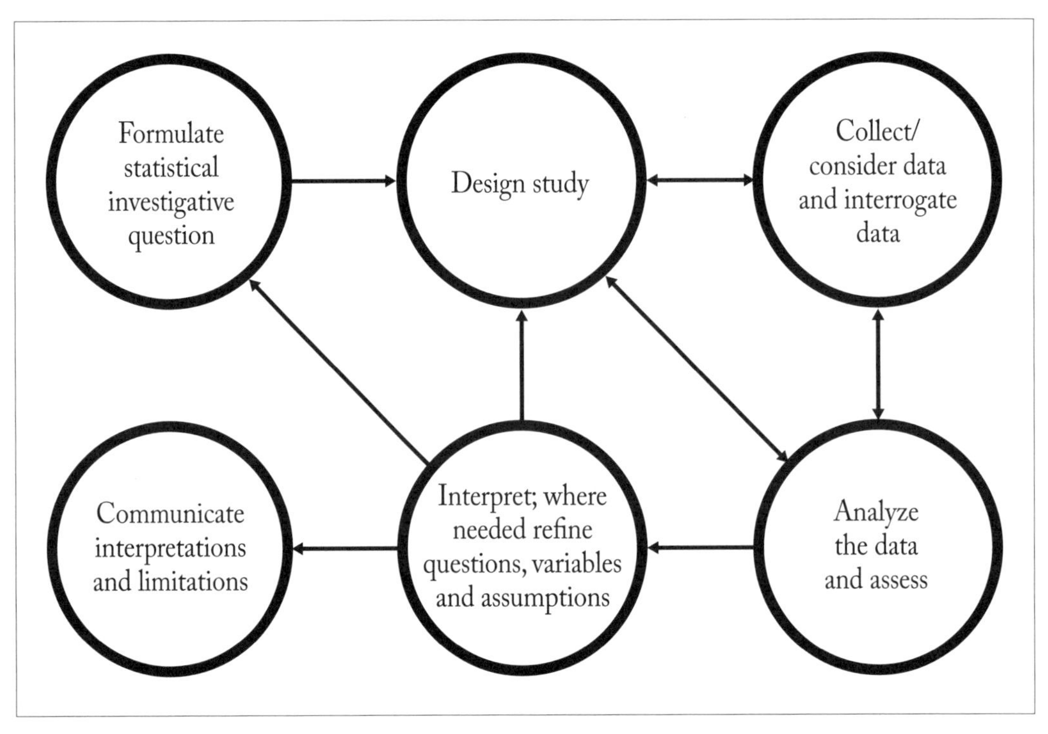

Source. The Pre-K-12 Guidelines for Assessment and Instruction in Statistics Education II (GAISE II) (Bargagliotti, A., Franklin, C., Arnold, P., Gould, R., Johnson, S., Perez, L., & Spangler, D. (2020). *Pre-K-12 Guidelines for Assessment and Instruction in Statistics Education (GAISE) report II.* American Statistical Association and National Council of Teachers of Mathematics).

variables, and building a first model. Using a statistical lens, these stages may also include the design of a study and data collection. The next phases, represented on the bottom row, represent opportunities to test, iterate, and create explanations using the model and its output. In mathematics, this may mean revising a function- or geometry-based model for a situation, engaging in interpolation or extrapolation, generating representations of the model, and crafting explanations that describe the model's efficacy in quantifying and predicting the contextual situation. This work is very similar in statistics—iterating models may mean changing assumptions and inputs, running multiple simulations, creating data displays, and writing about inferences, predictions, and conclusions.

Modeling Is More Than a Word Problem

Modeling, in this context, transcends the act of simply executing mathematical procedures on abstract or artificial problems, which can often feel disconnected from the real world and, thus, unrealistic or inaccessible for many students. Instead, it encompasses the vital skill of taking mathematical concepts and tools and using them to solve complex, real-life problems that carry practical significance. This approach not only deepens students' understanding of the subject but also showcases the utility and importance of mathematics in addressing the challenges and intricacies of the world they live in. By giving due consideration to the role of modeling, we can help students recognize the tangible benefits of mathematics beyond the classroom and empower them to apply their mathematical knowledge in meaningful ways throughout their lives.

When we consider applying mathematical concepts, a shift in perspective toward modeling is imperative. Modeling is a frequently misunderstood aspect of mathematics and differs from contextual and application problems (see Figure 2.3). To truly infuse relevance into the classroom and bridge the gap between mathematics and the real world, students need to grapple with complex and meaningful situations. It is within this complexity and relevance that mathematics becomes not just understandable but also valuable. Without these connections, mathematics may continue to be seen as just another subject that students temporarily learn to take a test, earn credit, or graduate from high school.

Figure 2.3
Contextual Problem Versus Modeling Problem

Typical Application Problem	Reestablishing Relevance (Modeling) Problem
A farmer has a field in the shape of a rectangle with a length of 100 m and a width of 50 m. She wants to build a fence around the perimeter of the field to keep her livestock secure. If the cost of fencing is $5 per meter, how much will the farmer spend on fencing?	The farmer wants to optimize the shape of her field enclosure to minimize fencing costs while ensuring the safety of her livestock. She needs to determine the dimensions and shape of the enclosure given certain constraints, such as total area, perimeter, and shape. Additionally, she must consider factors like the type of livestock, the animals' behavior, and any environmental conditions that might affect the fencing requirements. How can the farmer model this problem mathematically to find the most cost-effective solution for enclosing her livestock?

Note. Adapted from Garfunkel, S., & Montgomery, M. (Eds.). (2019). *GAIMME: Guidelines for Assessment and Instruction in Mathematical Modeling Education* (2nd ed.). Consortium for Mathematics and Its Applications (COMAP) and Society for Industrial and Applied Mathematics (SIAM).

Implementing this approach represents a significant departure from the conventional methods employed in high school mathematics. One thing to consider is that much of what has traditionally been taught in high school considers relationships between two variables. Modeling decisions, however, are often based on multiple factors, and we must pay attention to what new content students will need. To ensure that this happens, a comprehensive toolkit is needed that includes state standard-aligned curricula that seamlessly integrate relevant and meaningful modeling tasks, as well as professional learning opportunities that support this teaching approach.

Presently, educational materials incorporate application problems that may lack practical relevance or occasionally introduce authentic, real-world tasks but often relegate them to the conclusion of a chapter. The intent is to allow students to apply their newfound knowledge to these situations, which is a commendable initial step. However, this approach does not fully generate the necessary authenticity and relevance inherent in true mathematical modeling, which is essential for allowing students to experience the utility of mathematics.

A more engaging approach involves introducing these complex tasks at the outset of the instruction, coinciding with the development of mathematical concepts and procedures. This enables students to engage with the intricacies of the context and the concepts needed to navigate toward a solution. Such a pedagogical approach fosters a deeper understanding and appreciation of mathematics by making it an integral and immediately relevant part of the learning experience.

Modeling in the Classroom

Classroom instruction should engage students by helping them develop statistical and mathematical habits of mind. Engaging students in mathematical modeling allows them to see the relevance and usefulness of the concepts they are learning. Utilizing technology effectively works to build understanding and application of mathematics.

Consistent use of the eight Mathematics Teaching Practices from *Principles to Action: Ensuring Mathematical Success for All* (NCTM, 2014) provides increased opportunities for all students to actively engage in reasoning and making sense of the concepts. Through rich tasks, increased classroom discourse, and utilizing multiple representations, students develop a deep understanding of mathematics, allowing them opportunities to use those concepts to solve problems.

Preparing high school graduates for a technologically changing society with the ability to define and develop questions, think critically, reason with mathematics, and adapt to novel situations is our goal, as we stated in Chapter 1. Modeling provides a foundational structure within mathematics and statistics coursework in high school to promote the questioning, reasoning, problem-solving, and use and connection of mathematical concepts and representations to build this preparedness. Modeling also provides a setting to facilitate meaningful mathematical discourse, elicit clear evidence of student thinking, and demand productive struggle from students.

As teachers of mathematics, we can use modeling scenarios to motivate and support the learning of new mathematics as well as apply previously learned mathematics in new contexts. When students model, every expression, equation, and solution has meaning and must be interpreted by them in the context of the problem. This process of using mathematics to create a mathematical model that captures some essential aspects of some real-life process utilizes many of the effective teaching practices (NCTM, 2014). (See Figure 2.4.)

Figure 2.4
The Mathematics Teaching Framework

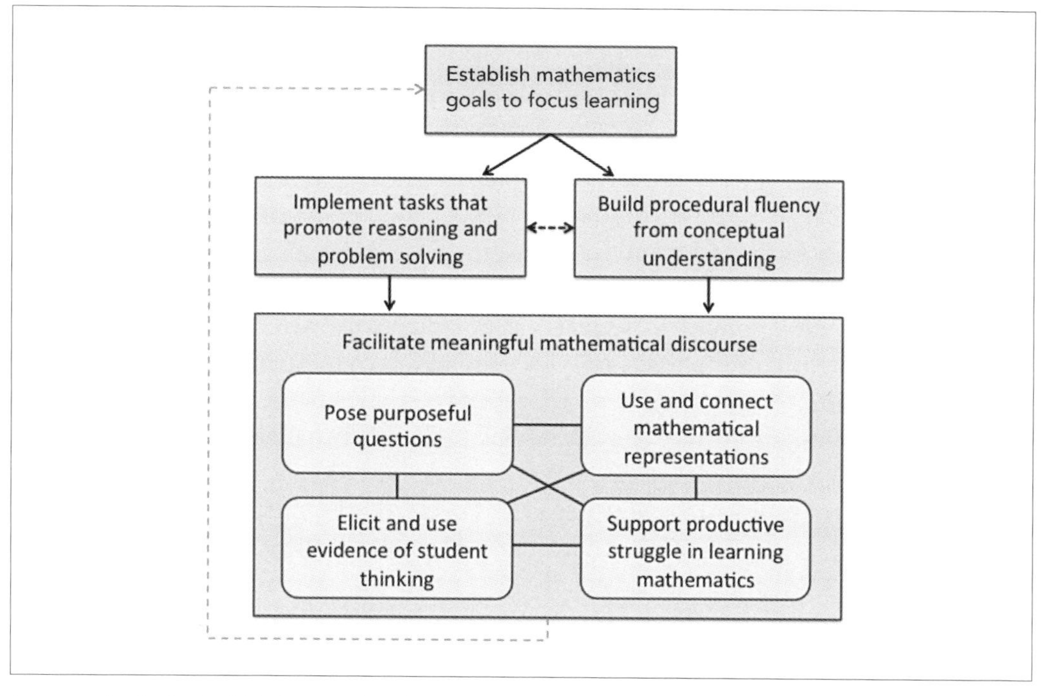

Source. Smith, M., Steele, M., & Raith, M. (2017). *Taking action: Implementing effective mathematics teaching practices grades 6–8.* NCTM.

Modeling tasks support and incentivize student decision-making. Rather than remembering what to do, the students *must* decide what to do, how to do it, and what the results of their modeling investigation represent within the context. These decisions require conversation and sharing of ideas along with mathematical and life experiences as students try different approaches and evaluate them in the light of their understanding of the processes being modeled.

Mathematical modeling in the classroom, as described by Rachel Levy (2015) in her short article "5 Reasons to Teach Mathematical Modeling," allows students to approach problems in their own way and generate a variety of useful solutions, thereby illustrating the creative nature of modeling and seeing the relevance of math in real-world contexts (which answers the ever-present student question, "Why do I need to know this?"). It also encourages teamwork and the sharing of ideas and methods.

Ownership of the mathematics occurs when students have the flexibility to make decisions about what to solve and how to solve it. Students are thinking their way through problems rather than just remembering what they were told to do or mimicking the teacher's steps. *Ownership is power, and power generates agency.* Students are engaged in creating their own problems or, at the very least, helping to frame problems associated with situations they encounter.

The students' experience in modeling must serve a course-specific pedagogical purpose beyond the modeling experience itself. The problem posed should, in some essential way, support the mathematics being studied.

Modeling Community and Social Contexts

Mathematics through modeling also provides students with tools to ask questions and develop solutions related to their communities and other larger social contexts. This approach seeks to empower students by looking at the communities and contexts they live in, asking mathematical questions, and then using mathematical and statistical modeling to understand and communicate about their communities. By aligning mathematical concepts and problem-solving skills with issues that matter to students and their communities, we foster the utility of mathematics and begin to equip students with the agency and motivation to leverage mathematics as a force for understanding, communicating, and addressing real-world challenges they experience. Mathematics teaching, learning, and application relevant to students' communities and cultures (Berry et al., 2020; Matthews & Banks, 2022) serve as compelling evidence that there is a growing demand for this pedagogical approach that emphasizes problem-solving through the lens of mathematics.

The range of modeling problems is vast and could include topics such as the following (see Appendix A for additional ideas):

- considering electoral district boundaries with constituent representation
- access to services in relation to income or demographics, including
  - healthy food, groceries, and supermarkets
  - air and water quality
  - noise
  - government services
  - green spaces
  - health care and emergency care
  - daycare
  - high-speed internet
  - sports and exercise opportunities
- considering the impacts of climate on communities
  - rain and flooding
  - heat, heat retention, and shade
  - other climatic factors
- youth sports and sports-related injuries, including concussions
- impact of traffic and urban planning
- social media usage and influence
- the varied impact of COVID-19 on communities

Consider the EV Charging problem that follows as an example in which students can consider using functions to describe data as well as piecewise functions.

The EV Charging Problem: When and How Much?

With the growth and expansion of electric vehicles (EVs), EV users must now consider their strategies for recharging their vehicles when they take a long trip. Unlike gasoline- or diesel-powered vehicles that refuel by refilling a tank with liquid, EVs require the battery on the vehicle to acquire additional charges. Drivers need to account for the time it takes for that process to happen.

Drivers can consider the relationship between the current charge of the battery, measured as the percent of the total charge, and the amount of time it would take to increase the charge to 100%. The Tesla Charge Time Calculator (https://evadept.com/calc/tesla-charge-time-calculator) is a great tool for considering this relationship. The data in Figure 2.5 were for a Model X P90D and Electrify America-Level 3-50 kW charger.

Figure 2.5
Starting Charge Level and Time to Charge 10%

Starting % of Charge	0	12	37	50	65	72	78	82	87	92	100
Time to 100% Charge (minutes)	221	208	181	167	151	144	137	121	88	54	0

Can you find an appropriate function to model this relationship?

The graph is not linear. The students have to grapple with areas of the data that look linear and areas that are nonlinear and must make some important decisions. After some discussion in their groups, supported by some thoughtful questioning from the teacher, students will typically convince themselves that their task is to determine a good strategy for recharging your EV if you are taking a long trip that will require multiple charging stops. Students may ask if the value in the distance that can be driven changes depending on how charged the car is. It turns out that a 2% charge consistently gets you 5 miles.

The students may realize that the data seem to be broken into three ranges: roughly 0 to 65, 65 to 82, and 82 to 90, with the most interesting section being the middle section. This process of culling the data is one of the central aspects of modern data science. Determining what aspects of the data are relevant to the problem under consideration is an increasingly important component of modern statistics education.

Some students may recognize that the data are, in fact, representative of a piecewise defined function. They have studied these in class, but this provides an example of how that algebraic structure is used outside of the classroom. In addition, students could expand their models to consider the additional time commitments for extra stops as compared with the costs and benefits of charging longer at stops. We have considered variations of this context, including charging cell phones and other items, as well as their power usage. Students could also see how different chargers at various charging opportunities might need to change their strategy for the long drive.

Modeling and statistical reasoning tasks play a vital role in bringing mathematical concepts to life for students. Students need opportunities to engage with these types of tasks alongside other tasks that support their development of mathematical concepts. Instead of attempting to find numerous modeling or statistical reasoning tasks that align with a single unit of study, consider a more sustainable approach that could involve integrating one of these types of tasks throughout a unit as an initial step.

CHAPTER 3

Reimagining Mathematics Content Connections

Can you think of a body of mathematics content that is memorable to students—in a positive way? If you are struggling or thinking "none of it," this is why we need to reimagine mathematics content. Students, for the most part, experience high school mathematics as a wide range of disconnected skills and procedures that they memorize, and they do not often see connections between what they have been required to learn and anything outside of the mathematics classroom. In 1995, the Third International Mathematics and Science Study, an international comparative study, characterized the United States mathematics curriculum as "a mile wide and an inch deep" (Schmidt et al., 1996, 1999). Although changes have been made to K–8 mathematics, little has changed since this time to mathematics in high school.

To aid students in seeing connections among topics, we offer a radical reconsideration of organizing the content of high school mathematics. This work emerges from the study of standards and curriculum over the past three decades to identify what the most important mathematical ideas are that thread throughout and transcend high school mathematics.

Further, as we recognized the importance of statistics, data science, modeling, and other critical content, content standards have continued to be added to high school mathematics. Standards efforts since that time have sought to shift that characterization, grouping aspects of mathematics content into larger bins in an effort to focus on bigger mathematical and statistical ideas. Although these efforts may have created incremental change, they have not resulted in a meaningful transformation of the high school mathematics landscape in ways that respond to the criticisms levied by Schmidt and colleagues, among others.

PAUSE AND REFLECT: Considering the Relevance and Utility of Mathematics Content

- How is high school mathematics content currently organized in your school/school system? What content currently gets the most focus? Why?
- What mathematics content does each and every student receive prior to graduation? How will this content prepare students for a technologically changing society?

- To what extent does the major high school mathematics content include an intentional focus on modeling and mathematical practices?
- Does your current high school mathematics content allow students to see the relevance and usefulness of what they are learning?
- How does or could the mathematics content, structure, and experiences contribute to the enduring practices envisioned for students?

Crosscutting Concepts

Five NCTM curriculum Crosscutting Concepts for high school mathematics provide a framework that builds on students' K–8 experiences, providing them with opportunities to make connections across areas of mathematics as well as organize their mathematics in ways that are useful for thinking about and questioning the world around them. These Crosscutting Concepts for high school mathematics provide a framework that builds on K–8 knowledge, allowing students to make sense of important mathematical concepts for the present and future. Although not grade-specific, the Crosscutting Concepts identify five overarching concepts that should be evident and developed throughout each high school course and that are critical to achieving vertical and horizontal coherence in high school mathematics.

Implementation of the high school Crosscutting Concepts will ensure that all students are provided with opportunities to make sense of important mathematical ideas and that the concepts we are teaching in mathematics classes are relevant and useful and speak to the enduring and connected understanding that every high school mathematics class must work to support students.

Crosscutting Concepts set the stage for a meaningful reorganization of high school course-taking opportunities. The Crosscutting Concepts are:

- Patterns and Generalization
- Variability and Change
- Functional and Structural Thinking
- Comparison, Difference, and Equivalence
- Making and Interpreting Predictions

The five high school Crosscutting Concepts serve as an organizing structure for planning and enacting high school mathematics lessons and courses. In the sections that follow, we define each of the Crosscutting Concepts by noting the important work that students do prior to high school and how high school mathematics should develop the Crosscutting Concepts; illustrate how the Crosscutting Concepts relate to the domains of algebraic and functional reasoning, statistical and data science reasoning, and geometric and measurement reasoning; and provide example tasks that center each Crosscutting Concept.

Patterns and Generalization

High school mathematics provides opportunities for students to identify patterns in abstract and modeling contexts, generalize those patterns with mathematical and statistical models, and make estimates and predictions based on those generalizations.

In middle school (NCTM, 2006), students make patterns and generalizations in a variety of contexts, including:

- **algebra:** represent, analyze, and generalize a variety of patterns with tables, graphs, words, and, when possible, symbolic rules;
- **geometry:** precisely describe, classify, and understand relationships among types of two- and three-dimensional objects using their defining properties;
- **data analysis and probability:** use observations about differences between two or more samples to make conjectures about the populations from which the samples were taken; and
- **data analysis and probability:** use proportionality and a basic understanding of probability to make and test conjectures about the results of experiments and simulations.

In high school, students will apply their understanding of patterns and generalization, building on what was established in prekindergarten through Grade 8, to a variety of modeling contexts and describe these patterns both recursively and explicitly. Students will analyze patterns in data to make generalizations, estimate solutions, and assess the reasonableness of predicted solutions.

Identifying and representing patterns is a foundational activity throughout mathematics (Table 3.1). Looking for patterns can occur across all representations rather than just within a single type of representation. When students are able to identify and describe patterns, they have the opportunity to ask why the pattern exists, which leads to the development of a solid understanding of the mathematical concepts being studied. Pattern recognition and generalization also apply to students looking at problems of interest, analyzing how their problem is similar or different to other problems they know how to solve, and ultimately increasing how they see the usefulness of the mathematics they are learning.

Table 3.1
Contexts for Patterns and Generalization Crosscutting Concept

Statistics and Data Science Contexts	Algebra and Functions Contexts	Geometry and Measurement Contexts
• Describing patterns and trends in distributions of data, including bivariate • Making inferences from sample data to a population • Looking for structures within data and developing models to represent data to make predictions	• Generating equations and inequalities to describe patterns, contexts, and/or data • Developing functional models to represent data	• Analyzing and comparing characteristics of two- and three-dimensional shapes • Identifying properties that hold between and within shapes • Developing equations and inequalities to represent patterns in figures

Examples of Questions for Patterns and Generalization

When you are working on the Patterns and Generalization Crosscutting Concept, here are some examples of questions that you can ask your students to center the idea:

- What patterns do you see in this data set, model, graph, equation, and so on?
- What type of model or representation could you use to represent this scenario, pattern, graph, table, equation, and so on?
- What would you expect the pattern to produce at Stage 20? Stage *n*?
- What conjectures can be made based on your thinking?
- How would you describe and communicate about the pattern and relationships using mathematical notation?

Variability and Change

High school mathematics provides opportunities for students to describe how quantities vary and covary, describe rates of change using mathematical and statistical language, create mathematical and statistical models that describe variability and change apparent in real-world phenomena, and construct inferences relative to variability and change based on their models.

In middle school (NCTM, 2006), students experience variability and change in a variety of contexts, including:

- **number and operations:** understand and use rates and proportions to represent quantitative relationships;
- **number and operations:** develop, analyze, and explain methods for solving problems involving proportions, such as scaling and finding equivalent ratios;
- **algebra:** explore relationships between symbolic expressions and graphs of lines, paying particular attention to the meaning of intercept and slope;
- **algebra:** use graphs to analyze the nature of changes in quantities in linear relationships;
- **measurement:** solve problems involving scale factors, using ratio and proportion; and
- **measurement:** solve simple problems involving rates and derived measurements.

Building on understandings established in prekindergarten through Grade 8, students had experience thinking about how two related quantities vary together. They examined linear and exponential relationships and described the ways in which those relationships change from data point to data point. Much of this experience focuses on constant rates of change.

High school mathematics can broaden and establish the overarching concepts of variability and change. Students can characterize rates of change more globally for a function or relation and describe the ways in which rates of change vary within and between mathematical relationships. Similarity, from a geometric or algebraic standpoint, describes change as a figure or function dilated by a scale factor. Trigonometric ratios represent specific rate-of-change relationships that are formalized either in triangles or as general functions. At its core the study of statistics is about understanding variability with standard deviation and variance as ways to quantify them within a data set.

Building from student understanding of ratio and proportional relationships, functions, and expressions and equations, students explore variability and change to understand the relationship between a model and the rate of change represented within the model. Students will make inferences on the type of model based on the variability and change. (See Table 3.2.)

Table 3.2
Contexts for Variability and Change Crosscutting Concept

Statistics and Data Science Contexts	Algebra and Functions Contexts	Geometry and Measurement Contexts
• Making decisions based on data requires understanding, explaining, and quantifying variability • Interpreting the predictability of the slope of a line of best fit within a context	• Estimating average rates of change • Developing relationships among distance, velocity, acceleration, and associated rates of change	• Calculating scale for geometric figures • Adapting the ratio for scale for use in linear, planar, and solid contexts, including using trigonometric ratios

Identifying and describing how variables and quantities, and their associated relationships, vary and change are critical components of high school mathematics. When students describe variability and change, they develop a deeper understanding of what the quantities represent and their interrelationships within their context. Considering, recognizing, and describing variability and change provides a lens and associated language for students to accurately describe a wide range of nonstatic real-world situations and contexts.

Examples of Questions for Variability and Change

When you are working on the Variability and Change Crosscutting Concept, here are some examples of questions that you can ask your students to center the idea:

- How would you find the average rate of change for a relationship that is linear? Exponential? Over a specific interval?
- How do rates of change of different function families compare?

- When you consider a data set and related statistics, how would you differentiate between variability within a group, variability between groups, and sample-to-sample variability?
- How would you describe how your statistical problem-solving and decision-making are based on understanding, explaining, and quantifying variability in the data within your context?

Functional and Structural Thinking

High school mathematics provides opportunities for students to make decisions on how to mathematize situations and data sets. The decision to describe a context/data set by a specific function or distribution enables students to access a range of mathematical tools and sets of related questions but also reduces the ability to answer other questions. The selection and application of functional and structural thinking influence modeling, analysis, and the process of making sense of a behavior, phenomenon, or set of data.

Prior to high school, many students had experience with decontextualizing and recontextualizing mathematical situations. This work may have included the use of variable expressions, equations, and functions to represent mathematical situations, measures of center and spread that summarize data sets and describe their variability, and experience with understanding and translating between measurement units.

In middle school (NCTM, 2006), students use functional and structural thinking in a variety of contexts, including:

- **algebra:** relate and compare different forms of representations for a relationship;
- **algebra:** identify functions as linear or nonlinear and contrast their properties from tables, graphs, or equations;
- **algebra:** use symbolic algebra to represent situations and to solve problems, especially those that involve linear relationships;
- **algebra:** model and solve contextualized problems using various representations, such as graphs, tables, and equations;
- **geometry:** use geometric models to represent and explain numerical and geometric relationships; and
- **data analysis and probability:** make conjectures about possible relationships between two characteristics of a sample on the basis of scatterplots of the data and approximate lines of fit.

Building on their understanding of expressions, equations, and functions, students will explore function families to analyze key functional behavior, including end behavior, maxima/minima, intervals increasing/decreasing, and points of inflection. Students will develop and apply their understanding of function families to model and describe real-world contexts and/or data and develop analysis skills to assess the reasonableness of the models. In statistics and data science, the application and analysis of distributions allow students to build beyond measures of center and spread toward more nuanced descriptions of a data set's behavior, including linking features of the data set to the context from which it is drawn. (See Table 3.3.)

Table 3.3
Contexts for Functional and Structural Thinking Crosscutting Concept

Statistics and Data Science Contexts	Algebra and Functions Contexts
• Generating a functional equation of best fit (the structure) to model a data set, taking into account variability	• Describing functional behavior within a functional family

Developing a deeper understanding of functional behavior is an important part of high school mathematics. As more families of functions are learned, students have the opportunity to select a family to describe contextual relationships. Students begin to recognize that determining which function is used to model a scenario has implications for the types of questions they might ask and the features they might investigate. Similar opportunities and constraints follow as students consider and select distributions to apply to data.

The development of structural differences could be applied to geometry in the broader sense (Euclidean, spherical, and other non-Euclidean structures or relational and axiomatic approaches); however, the current application of geometry does not lend itself to consideration of these structures without significant curricular changes. It is unclear if these efforts are warranted, given the student-centered goals for high school mathematics.

Examples of Questions for Functional and Structural Thinking

When you are working on the Functional and Structural Thinking Crosscutting Concept, here are some examples of questions that you can ask your students to center the idea:

- What is a function family? What are the key features or graphical behavior of a given function family (linear, exponential, quadratic, trigonometric, etc.)?
- How can you use key features or graphical behavior to determine what type of function best models a given scenario, table of values, or graph of a relation?
- What potential limitations do function models have for a given data set and/or context?

Comparison, Difference, and Equivalence

High school mathematics provides opportunities for students to compare situations using mathematical and statistical models, to create multiple mathematical and statistical models for a phenomenon and compare their explanatory power, and to use geometric principles to identify similarities and differences in real-world situations. Students use equivalent forms of algebraic and geometric representations to reveal aspects of structure.

In middle school (NCTM, 2006), students use comparison, difference, and equivalence in a variety of contexts, including:

- **number and operations:** compare and order fractions, decimals, and percents efficiently and find their approximate location on a number line;
- **number and operations:** develop meaning for integers and represent and compare quantities with them;
- **number and operations:** develop an understanding of large numbers and recognize and appropriately use exponential, scientific, and calculator notation;
- **number and operations:** use the associative and commutative properties of addition and multiplication and the distributive property of multiplication over addition to simplify computations with integers, fractions, and decimals;
- **number and operations:** develop, analyze, and explain methods for solving problems involving proportions, such as scaling and finding equivalent ratios;
- **algebra:** relate and compare different forms of representations for a relationship;
- **algebra:** recognize and generate equivalent forms for simple algebraic expressions and solve linear equations;
- **geometry:** create and critique inductive and deductive arguments concerning geometric ideas and relationships, such as congruence, similarity, and the Pythagorean relationship;
- **geometry:** examine the congruence, similarity, and line or rotational symmetry of objects using transformations;
- **measurement:** understand the relationship among units and convert from one unit to another within the same system; and
- **data analysis and probability:** find, use, and interpret measures of center and spread, including mean and interquartile range.

Comparison is central to better understanding quantities, expressions, structures, families of functions, and distributions. The foundational activity of sorting and clustering provides a tool for recognizing the similarities or attributes of a structure and defining what makes it different from others. This is key for students to develop their understanding of and ability to choose and apply various mathematical structures.

Equivalence is the foundation of algebraic manipulation and supports students in moving between forms to reveal key features of a mathematical phenomenon and generate solutions to algebraic questions. Recognizing and understanding equivalent structures as well as differing structures contributes to students' better understanding of the mathematical domains they are learning. These absolutes also contribute to the development of similarity, which provides consistency with certain parameters and differences, often structured, in other parameters.

Comparison through determining equivalence or differences is fundamental to mathematics (Table 3.4). Although comparison and equivalence could be considered separately, there are benefits to developing them together. Equivalence can be used to highlight different features of relationships.

Table 3.4
Contexts for Comparison, Differences, and Equivalence Crosscutting Concept

Statistics and Data Science Contexts	Algebra and Functions Contexts	Geometry and Measurement Contexts
• Comparing distributions of variables with respect to variability and measures of what is typical and plausible • Comparing lines/curves of best fit for bivariate data • Designing studies to compare two or more groups for statistical differences in their distributions	• Comparing how and where different functions support and do not support a specific context • Comparing function families in regards to behaviors and qualities • Analyzing and comparing the utility of equivalent forms of equations	• Comparing angles and figures • Comparing transformed figures to their preimage • Identifying congruence of angles or figures and as a 1:1 case of similarity

Examples of Questions for Comparison, Difference, and Equivalence

When you are working on the Comparison, Difference, and Equivalence Crosscutting Concept, here are some examples of questions that you can ask your students to center the idea:

- What information does this form of the equation best highlight?
- How drastically could distributions with equivalent statistical measures differ?
- Is this graph a fair representation of the data? What parts might be misleading?

Making and Interpreting Predictions

High school mathematics provides opportunities for students to use models to predict the behavior of a real-world situation to identify constraints and limitations to a model and its predictive ability, vary the constraints of models and the impact on its predictability, and assess the reasonableness and efficacy of mathematical and statistical techniques for making, analyzing, and predicting current and future behavior.

In middle school (NCTM, 2006), students use making and interpreting predictions in a variety of contexts, including:

- **number and operations:** develop and use strategies to estimate the results of rational-number computations and judge the reasonableness of the results;
- **number and operations:** develop, analyze, and explain methods for solving problems involving proportions, such as scaling and finding equivalent ratios;

- **algebra:** model and solve contextualized problems using various representations, such as graphs, tables, and equations;
- **geometry:** use geometric models to represent and explain numerical and algebraic relationships;
- **data analysis and probability:** formulate questions, design studies, and collect data about a characteristic shared by two populations or different characteristics within one population; and
- **data analysis and probability:** use proportionality and a basic understanding of probability to make and test conjectures about the results of experiments and simulations.

Entering high school, students may have explored real-world situations, identified dependent and independent variables, and engaged in interpolations and extrapolations. They also may have made and interpreted predictions by extending patterns, developing mathematical models, or using basic principles of probability.

Making and interpreting predictions in high school focuses on creating and interpreting mathematical models. Predictability questions arise as students describe the nature of a relationship that is being modeled, identify constraints and limitations to the model, and assess the model's predictive ability. This work includes lines and curves of best fit and regression in statistics, properties of shapes and similarity in geometry, and families of functions. Students assess the reasonableness of solutions, including the appropriateness and limitations of the proposed models.

Working with making and interpreting predictions is important in mathematics (Table 3.5). Students often recognize the usefulness of analyzing predictability. They recognize that modeling situations involve error and begin to develop tools to analyze and describe the error. The development of predictability with error provides a more realistic portrayal of the usefulness of mathematics when applied in the real world. Additionally, students can use probabilities to help make informed decisions.

Table 3.5
Contexts for Making and Interpreting Predictions Crosscutting Concept

Statistics and Data Science Contexts	Algebra and Functions Contexts	Geometry and Measurement Contexts
• Assessing the reasonableness of a model and estimating errors in conclusions • Using such measures as margin of error to quantify variability to be more precise with predictions • Estimating probabilities and using probabilities to evaluate outcomes of decisions	• Modeling a scenario with a function and determining the reasonableness and/or limitations of the model	• Determining reasonableness for measurement changes between a preimage and an image

Examples of Questions for Making and Interpreting Predictions

When you are working on Making and Interpreting Predictions Crosscutting Concept, here are some examples of questions that you can ask your students to center the idea:

- What is an acceptable range for error given a set of predictions? Should the range of errors be all the same? Why?
- What parameters could you change to limit the range of error?
- Which parameter most affects the range of error? Which least affects the range of error?

The Crosscutting Concepts should be woven throughout all courses in high school and described through both a mathematical and statistical lens. They should be visible to students, and it is incumbent upon us to explicitly connect the Crosscutting Concepts to the specific mathematics content under discussion with our students. As students make choices about the specific mathematics and statistics content they wish to explore through course selection later in high school, the Crosscutting Concepts should remain a common theme. For example, a student taking AP Calculus BC, a student taking a fourth-year quantitative reasoning course, and a student taking a data science course should all be able to identify how the specific content in their courses relates to patterns and generalization, rates of change, functional and structural thinking, comparison, equivalence and difference, and predictability. As districts and schools organize courses and units, development should be grounded in the Crosscutting Concepts.

With a focus on reasoning, the high school Crosscutting Concepts are designed to engage students in actively doing and making sense of mathematics and actively applying mathematics to understand the world around them. These activities are designed to work to positively transform and enhance the enduring mathematics learning of students.

Organizing Content Into Courses

We will explore practical options for organizing the content within the Crosscutting Concepts to make high school mathematics relevant and meaningful for students. The high school Crosscutting Concepts outline the connections among topics that educators should be developing for all students throughout each high school course.

Catalyzing Change in High School Mathematics (NCTM, 2018) states that implementing a high school mathematics curriculum focused on the Essential Concepts and approaching those Essential Concepts with rigor and equitable instructional practices can increase the likelihood of every student taking high school mathematics courses that foster students' development of deeper understanding of mathematics, support their ability to make sense of the world, and increase their opportunities.

(See Appendix B of this book for more information on the connections between Crosscutting Concepts and the Essential Concepts.)

High school curricula should be coherent both vertically and horizontally to help ensure success for students. Strong course alignment will help ensure that students build upon their mathematical knowledge progressively and that the students see that concepts taught in mathematics classes are connected, relevant, and useful.

PAUSE AND REFLECT: Considering Course Options and Implications

- Think about your school or district's current high school mathematics course options. To what extent do the mathematics courses prepare and support your students to see a problem and critically think about ways to respond?
- Do the courses provide your students with mathematical learning experiences that evoke creativity and a sense of curiosity?
- Do the courses provide experiences that help students see the relevance and usefulness of mathematics?
- Do the courses allow students to see themselves as proficient in understanding and learning mathematics and able to apply it to a range of their interests?
- To what extent do the courses prepare your students for an unknown future?

Criteria for Reimagining

As we reimagine high school course content and instructional practices, there are several factors to consider. The criteria that follow compel us to consider how the systemic structure of high school mathematics must adapt. We must address these factors so that we are better supported and empowered to innovate instruction and learning opportunities that are designed to create dynamic, engaging, and enduring experiences for students.

Optimal Duration of Mathematics Studies

Catalyzing Change in High School Mathematics (NCTM, 2018) recommends 4 years of continuous engagement in the study of mathematics at the high school level. Studying mathematics each year of high school ensures steady and continuous growth toward future opportunities. Compressing mathematics into fewer than 4 years, or allowing gaps of more than 1 year, interrupts the learning cycle and limits continued growth and opportunity (NCTM, 2018, p. 5). Effective design of high school mathematics pathways includes opportunities for all students to experience common coursework and then explore robust pathway options over 4 years.

Even though not all states and districts require 4 years for graduation requirements, the 4 years of continual mathematics learning in high school is critical for student success in future endeavors. All other things being equal, students taking math in their senior year have a 15% higher chance of completing a 4-year degree than those who do not (Kung, 2023). These 4 years must be relevant and engaging for students rather than merely taking courses with external positive or detrimental effects. Providing interest-driven pathways increases opportunities for students to seek out mathematics courses beyond what may be required.

Interest-Driven Pathways Should Increase Opportunities

Students should have a common shared pathway for the first 2 or 3 years of high school. The final 1 or 2 years of high school mathematics might offer a variety of options based on student interest.

Mathematics interest-driven pathways should support students toward their future goals beyond high school. They must be relevant and customized to meet students' specific interests and goals. States and districts must deliberately plan to ensure the successful implementation of relevant, personalized mathematics pathways for success.

Creating a meaningful pathway to high-level mathematics for all students, including calculus, is an important part of any revitalization design. Ideally, we want to create a system for K–12 mathematics where more students choose to pursue the study of advanced mathematics, allowing and inspiring them to pursue STEM fields of study and careers. However, the system should not limit students to a single pathway only, allowing them to deviate if they are deemed unworthy or incapable based on an individual or a standardized test result. All pathways need to be aligned with higher education admissions and degree requirements. Options designed to prepare students to succeed in STEM fields, such as data science and non-STEM studies, should be included, as well as those that prepare students to succeed in career and technical education coursework and high-value certificates in the trades.

Students should also be developing a deeper understanding of concepts that are applicable to their interests rather than merely completing coursework to allow them to register for the next course. For example, calculus is not required for admission outside of specific majors (Anderson & Burdman, 2022), and high schools should provide a variety of interest-driven pathways to better engage and meet the needs of students. Mathematics leaders at the local and state levels and from high schools and postsecondary institutions should collaborate to ensure smooth transitions from high school to and through community college degrees and programs and to 4-year degrees.

The University of Texas Charles A. Dana Center, through the Dana Center Mathematics Pathways, has been working to create pathway change in mathematics by addressing systemic strategies and involving a variety of stakeholders from national associations to the classroom levels. The Dana Center Mathematics Pathways (The University of Texas Charles A. Dana Center, 2019) has defined its mathematics pathways in the form of these four principles:

- are aligned to student goal
- accelerate student progress towards completion

- integrate student learning supports
- use evidence-based curriculum and pedagogy

Mathematics as a Gatekeeper

Mathematics, unfortunately, has been used for many years as a gatekeeper to advanced coursework and has denied students access to a multitude of career pathways. College algebra traditionally has been the default math class for most students in the United States. Recent studies have shown that college algebra has served as a mathematical gatekeeper and is irrelevant for many postsecondary pathways, creating major equity issues and limiting access to relevant mathematics coursework for various career fields (The University of Texas Charles A. Dana Center, 2019, The Launch Years). According to Gordon (2008), "Eighty percent of students do not need an algebra-intensive curriculum, nor calculus, to succeed in their degree programs." Decisions relating to mathematics pathways, and the elitist position that tracks to calculus have taken, particularly affect underserved, Black, and Latino students (Jimenez et al., 2016; EdSource, 2012; Burdman, 2018).

Secondary mathematics courses must prepare students for their postsecondary pursuits, and as such, we need to design course pathways as bridge-building endeavors. Viable course pathways must ensure that throughout students' high school mathematics journeys, their courses will emphasize key mathematical and statistical understanding and practices, provide equitable opportunities to access high-quality teaching, and engage students in learning experiences that help them to develop important habits of mind for seeing, understanding, and using mathematics and statistics relative to their daily lives, their interests, and matters within and beyond their communities (NCTM, 2018). Districts must continually examine student experiences in each interest-driven pathway to ensure that students are developing the necessary skills and understanding to be successful in future endeavors.

High Expectations for *All* Students

We believe that all students have the ability to learn high-level mathematics. Students in special populations in schools are often not given the same opportunities to learn mathematics as their peers. Students with disabilities can learn and understand mathematics when they are placed in a system that has beliefs in their abilities, as well as the understanding that not all students learn in the same way. Preparing for and allowing for differentiation is key to helping students succeed in the mathematics courses that fulfill their selected pathway. The same is true for multilingual learners who have the cognitive ability to learn mathematics and are challenged to learn both mathematics and a new language simultaneously. Supporting learners' language development through the use of language routines provides them with the structures necessary to successfully do both.

It is important to note that all courses and pathways offered in high school should be open and accessible to any student interested. Although not all students will require advanced levels of statistics or calculus to satisfy their future majors or careers of interest, some students will want to take courses beyond the 4 years of mathematics required. Pathways should be designed to cultivate students' interests and prepare them with the mathematics content and understanding, even beyond the minimum requirements.

Acceleration and Tracking

Accelerating students may be appropriate when they have demonstrated a deep understanding of the mathematics standards beyond their current levels. No critical concepts should be rushed or skipped (NCTM, 2018). Students who are rushed through content are often those who drop out of mathematics when given opportunities to do so (Boaler, 2016).

Tracking is based on perceived student ability, race, socioeconomic status, gender, language, or other expectations determined by adults (Stiff & Johnson, 2011). It puts students into courses with drastically differing access to quality mathematics instruction to prepare them for success in postsecondary opportunities. Those in "low" tracks often focus on rote procedures with little emphasis on developing understanding (Oakes, 1985). Students placed in a "high" track often experience instruction that focuses on developing conceptual understanding, problem-solving, and thinking skills.

Interest-driven pathways are different from tracking. Interest-driven pathways are based on student interest rather than on perceived ability. All pathways must focus on reasoning and sense making so that future opportunities are not limited. Although interest-driven pathways can accommodate acceleration, they must not become de facto tracking where the quality of instruction, the expectations of students, and opportunities for students are limited and prescribed.

This work of transforming high school mathematics experiences is difficult and involves many factors. Differing structures among states within K–12 and 3-year and 4-year higher education colleges and universities add to the complexity; what works in one state often does not easily translate to another due to these differing structures. Work to remake a system in which mathematics has been a gatekeeper is challenging yet necessary to increase opportunities for every student. As districts address this challenge, it is vital to engage the school and district leaders, beyond just the mathematics leaders and the local community, from the beginning to help ensure understanding and successful implementation.

Following are examples of potential courses offered in the interest-driven pathways for 11th and 12th grades. Each option provides opportunities for students to engage in reasoning and sense making and allows students to see the relevance and utility of the concepts they are learning. The interest-driven courses build on prior work and provide additional connections to the Crosscutting Concepts.

Course Examples

Quantitative Reasoning

Students who are interested in careers such as those in social sciences, humanities, fine arts, and performing arts might take quantitative reasoning courses. Through quantitative reasoning, students further develop their range of mathematical and statistical concepts and skills and continue to develop their problem-solving strategies by engaging in real-world applications. Students utilize critical thinking skills as they question and analyze data, interpret graphs, and make informed decisions based on information presented in a range of formats. They employ mathematical and statistical modeling as they apply mathematical and statistical principles to analyze scenarios from various fields, such as economics, science, and social sciences, in settings that could include population dynamics, ecological systems, and the spread of infectious diseases.

Throughout the course, students develop and apply a range of mathematical tools in the areas of probability, statistics, algebraic modeling, and financial literacy. Variability and change are foundational in investigating scenarios ranging from analyzing trends in population growth to understanding the implications of statistical data in medical research to examining the impact of economic policies on personal finances. Students build on prior work from Making and Interpreting Predictions and other Crosscutting Concepts as they engage in assessing and quantifying risks in a variety of contexts, such as gambling and insurance. Throughout the course, students develop their ability to use and apply mathematics as they gain a deeper appreciation for the role of mathematics and statistics in questioning and addressing real-world issues and making informed decisions in their lives.

Statistics

Students who are interested in careers such as those in nursing, psychology, and social sciences might take statistics. A typical high school statistics course engages students as they continue to develop an understanding of the concepts and methods used to analyze and interpret data. Through hands-on activities, projects, and real-world examples, students learn how to ask statistical questions, collect and organize data, select and implement appropriate analysis tools, make sense of the results with respect to the context, and represent, summarize, and communicate the findings to answer the posed questions. Students delve into topics such as descriptive statistics, probability, and inferential statistics. The student experiences are oriented around developing their skill and ability to apply the statistical problem-solving cycle.

Probability is a component of a high school statistics curriculum, enabling students to build on prior understandings from the Making and Interpreting Predictions Crosscutting Concept as they further their ability to quantify uncertainty. Through simulations and experiments, students deepen their understanding of probabilistic outcomes and learn how to apply probability models to various real-life situations.

Students deepen their understanding of concepts in the Patterns and Generalization Crosscutting Concept as they draw conclusions and make inferences about populations based on sample data using inferential statistics. Students develop and apply their understanding of variability as they engage in a variety of situations as they learn hypothesis testing, confidence intervals, and principles of statistical inference as they analyze survey results, conduct experiments, and draw meaningful conclusions about the world around them. A high school statistics course equips students with valuable analytical skills and prepares them to critically evaluate data-driven arguments in a wide range of situations and scenarios.

Data Science

Students who are interested in careers such as business analysis, finance, marketing, and data science might take a data science course. A high school data science course integrates statistics, programming, and domain knowledge as students explore how large data can be used to make predictions in various fields and understand the importance of data-driven decision-making in the modern world. The course builds from an understanding of statistics and the statistical problem-solving cycle, leading to techniques and related questions when working with large data. Data science builds on the perception of data as numbers by recognizing that data can also be text, images, sound, and video.

Students delve into programming languages commonly used in data science. They learn how to ethically and effectively clean data sets, visualize data using a variety of representations, and ensure thorough analyses to generate predictions and accurate conclusions. Through coding experiences, students not only strengthen their programming skills but also deepen their understanding of how these tools can be applied to data analysis to address authentic domain problems.

Table 3.6, from the NCTM position statement *Teaching Data Science in High School: Enhancing Opportunities and Success*, shows the cognitive benefits to students as well as the habits of mind that can be developed.

This understanding is grounded in both the Crosscutting Concepts of Making and Interpreting Predictions and Variability and Change. Throughout the course, students develop their understanding and knowledge of predictive modeling. They learn about different algorithms and techniques used to train models and make predictions based on data patterns. Students apply their knowledge to tackle complex problems and gain a deeper appreciation for the power of data science in understanding the world around them.

Table 3.6
Cognitive Benefits and Habits of Mind for Students

Skills: How Students Engage Cognitively With Mathematics and Data
Students should do the following:
Use mathematics and data to make logical and informed decisions.
Interact with relevant contexts through rich and accessible data.
Reason deductively and inductively with data.
Formulate and test predictions based on finding, sorting, characterizing, and analyzing mathematical and statistical models.
Recognize which mathematical strategies and tools are efficient in a given data situation.
Develop flexible and creative problem-solving through data-driven processes.
Visualize, model, and construct multiple representations for authentic and data-rich situations while making connections among representations.
Justify conclusions and critique the reasoning of others through data investigations.
Communicate effectively and precisely through a data lens.
Tackle ethical and social issues through data collection/consideration, data analysis, and communication of results.
Work independently as well as in teams to ask meaningful questions and make logical and data-informed decisions.

Habits of Mind: The Dispositions That Shape a Student's Identity
Students should develop the habits of mind that allow them to develop the following mindsets:
Be willing to be wrong in search of the truth.
Be open to challenging questions and being challenged.
Exhibit curiosity about the stories in data and mathematical relationships.
Develop a mindset for persistence, for challenge, and for seeing failure as an opportunity to refine and elaborate.
Appreciate statistical/mathematical models as ways to answer questions and understand the underlying context of a problem or situation. Appreciate the meaningful attributes that data and mathematical models can show about a situation.
Believe that mathematics and statistics can be used meaningfully to make informed decisions.
Develop confidence to move from being data consumers to becoming data producers and analyzers.
Be willing to question, analyze, and challenge the accepted meaning of statistical and mathematical models.
Become risk-takers while engaging with relevant models and data.
Recognize the importance of understanding risk and its role in informed decision-making, knowing that every decision will have benefits and costs that need to be considered in making the decision.

Calculus

Students who are interested in STEM careers, including those in the physical sciences, engineering, mathematics, as well as economics, might take a calculus course. Typical courses often begin with a review of algebraic and trigonometric functions, as well as limits and continuity. Students may then begin developing concepts around differentiation as they build from prior work in the Variability and Change Crosscutting Concept when learning to represent rates of change, slopes of curves, and instantaneous rates of change at a given point. Integration and its applications are developed through learning a variety of techniques as students explore topics such as areas under curves, volumes of revolution, and solving problems involving accumulation. The connection between these central themes of calculus provides an opportunity to support students' development of concepts in the Functional and Structural Thinking Crosscutting Concept.

We encourage students to apply calculus concepts to real-world scenarios, fostering critical thinking and problem-solving skills. They may engage in a variety of challenging problems that require them to analyze, interpret, and communicate mathematical ideas effectively. By the end of the course, students should have a solid understanding of the principles of calculus and be equipped with the tools to solve complex mathematical problems, laying a strong foundation for further studies.

Since the release of *Catalyzing Change in High School Mathematics* (NCTM, 2018), many schools, districts, and states have made strides toward meeting the recommendations outlined in this work. Examples of notable efforts from states and national leaders can be found in supporting resources at nctm.org/reimagined. We must continue our work in developing and refining the high school mathematics experience that is relevant and engaging for our students.

Although different districts and states have created different models of pathways, they recognized the need to include representation from multiple stakeholders as the models were developed. Voices can include those from higher education, postsecondary career training, and the business sector, as interest-based pathways must work to expand, not limit, access to postsecondary options. Whether the models leverage modules, semester-long, or full-year courses, they all can open doors for students heading to the workforce, military, colleges, universities, and certificate programs.

CHAPTER 4

Revitalizing the Student Experience

How we engage students in learning mathematics impacts how students view mathematics and themselves. Do *we* ask them to make sense, question, and apply mathematics, or do *we* ask them to remember a set of made-up rules? Do *they* perceive the utility of mathematics? Do *they* see themselves as learners and users of mathematics? The answers to these questions are largely determined by the instructional practices that *you* employ in your classroom. Their short engagement with mathematics learning carries on long after they leave high school.

When we implement the eight Mathematics Teaching Practices from *Principles to Actions* (NCTM, 2014), students experience mathematics learning as a process of making sense of and using their mathematics, which in turn develops skills such as communication, collaboration, creativity, problem-solving, and adaptability. Integrating these teaching practices into mathematics courses better prepares students for a rapidly changing job market and society. As mathematics pathways are designed for students, leaders should anticipate the changing landscape of careers and industries. Mathematical and statistical reasoning and comprehension that leads to critical thinking and problem-solving are the skills needed for the jobs of the future (World Economic Forum, 2023).

Mathematical and Statistical Processes

High school mathematical study should develop in students the understanding of and ability to apply a foundation of mathematical and statistical processes needed for college and career success and the belief in their ability to understand, learn, and apply mathematics. Time and again, research has shown that students' self-efficacy related to mathematics—not specific learned aspects of mathematics content—has been a critical factor in students' postsecondary mathematics success (e.g., Bengmark et al., 2017; Czocher et al., 2020; Larson et al., 2015). The mathematical and statistical processes that students use to model and explore mathematics and statistics situations are key tools that make mathematics generative (Schoenfeld, 2020) and the mathematical tools students find useful in their futures. These processes include

- modeling and using tools and representations
- explaining, reasoning, and proving in mathematics and statistics
- seeing, describing, and generalizing structure
- developing the habits of a productive mathematical and statistical thinker

Table 4.1
Mathematical and Statistical Processes

Modeling and Using Tools and Representations
• Model with mathematics and statistics.
• Decontextualize and recontextualize mathematical and statistical situations.
• Use appropriate tools, including technology, strategically.
• Use representations to examine multiple mathematical and statistical points of view.
Explaining, Reasoning, and Proving
• Conjecture and reason inductively and deductively.
• Construct viable arguments and critique the reasoning of others.
Seeing, Describing, and Generalizing Structure
• Look for and make use of structure.
• Look for and express regularity in repeated reasoning.
Habits of a Productive Mathematical and Statistical Thinker
• Make sense of problems and persevere in solving them.
• Attend to precision in mathematical and statistical language and processes.
• Tinker productively with mathematical and statistical ideas and problems.

The 11 mathematical and statistical processes shown above in Table 4.1 form the enduring practices every student should have as they leave high school. They derive from decades of research on the productive mathematical and statistical habits of mind of professionals who use and apply mathematics every day (e.g., Bailey & McCulloch, 2023; Cuoco et al., 1996, 2010; Lim & Selden, 2009; Matsuura et al., 2013). They build meaningfully on the Standards for Mathematical Practice (CCSSO, 2010) consistent across many states. These processes outline the key proficiencies that students should develop and apply across lessons, units, and courses of study and that they should have internalized when they leave high school.

The 11 mathematical and statistical processes provide a framework for teachers to thoughtfully consider how, regardless of the mathematical content taught, they are supporting students to develop critical thinking and analysis skills to reason thoughtfully. Organizing the mathematical practices into the following categories allows educators to better see the connections between these processes. These processes, paired with a theme of mathematical modeling, should form a consistent structure for every high school mathematics course.

Students should regularly be developing these processes as they engage with understanding the content. Not all 11 will be utilized in each lesson, but it is crucial to provide adequate opportunities for students to engage with each of them consistently. Some content lends itself to some of the four process clusters more than others. As educators design courses, careful attention should be paid to integrating these practices throughout the student experience.

Following is a bulleted list of some key student experiences for each of the mathematical and statistical processes.

Modeling and Using Tools and Representations

- Model with mathematics and statistics.
    - Use mathematical and statistical tools to describe situations from everyday life, society, and the workplace.
    - Consider the effects of the assumptions made in analyzing situations.
    - Reflect on the results and iteratively improve the model as necessary.
- Decontextualize and recontextualize mathematical and statistical situations.
    - Represent mathematical and statistical situations symbolically.
    - Perform necessary calculations in analyzing symbolic representations.
    - Determine what makes sense for the context of the situation and make appropriate modifications.
- Use appropriate tools, including technology, strategically.
    - Contemplate and determine which tool is most useful for solving a mathematical or statistical problem.
    - Utilize a variety of tools to analyze and deepen understanding of concepts.
    - Consider benefits and limitations when determining when and what tools should be used.
- Use representations to examine multiple mathematical and statistical points of view.
    - Given a situation, make multiple conjectures and explore the validity of those conjectures using mathematics and statistics.
    - Identify relationships between various representations and their meaning within the mathematical and statistical situations.
    - Consider the value and limitations of specific mathematical and statistical representations and how they can contribute or obscure information in various settings.

Explaining, Reasoning, and Proving

- Conjecture and reason inductively and deductively.
    - Given a situation, make multiple conjectures and explore the validity of those conjectures using mathematics and statistics.
    - Develop mathematical or statistical arguments using techniques including cases, counterexamples, and visual representations.
    - Make justifiable predictions from data using mathematical and statistical models, taking into account context and other factors.

- Construct viable arguments and critique the reasoning of others.
    - > Utilize assumptions, definitions, and prior results in constructing arguments.
    - > Compare arguments and strategies and describe their strengths, limitations, and connections to the problem or mathematics and statistics being discussed.
    - > Question and productively challenge mathematical and statistical arguments based on data and reasoning to clarify and improve them.

Seeing, Describing, and Generalizing Structure

- Look for and make use of structure.
    - > Identify and describe patterns.
    - > Recognize complex structures, such as some expressions, can be viewed as single objects as well as being composed of multiple objects.
    - > Apply general mathematical or statistical rules to transform representations to reveal different characteristics or features.
- Look for and express regularity in repeated reasoning.
    - > Make generalizations when a repeated calculation occurs.
    - > Accurately describe the reasons behind mathematical procedures.
    - > Represent and describe mathematical and statistical patterns and phenomena.

Habits of a Productive Mathematical and Statistical Thinker

- Make sense of problems and persevere in solving them.
    - > Look for multiple entry points when solving mathematical and statistical problems.
    - > Analyze the information provided and the constraints of the problem to help determine a solution approach or strategy.
    - > Recognize that changing one's approach or strategy may be needed to generate a solution.
- Attend to precision in mathematical and statistical language and processes.
    - > Communicate precisely with others.
    - > Calculate accurately and efficiently.
    - > Connect informal and contextual language with precise definitions of mathematical and statistical concepts.
- Tinker productively with mathematical and statistical ideas and problems.
    - > Consider starting points, test cases, and boundary inputs that help interpret and make meaning of context as well as mathematical and statistical ideas.

- Consider how ideas, concepts, and similar problems relate to or differ from the concepts or problems being explored.
- Monitor progress, assess and evaluate strategies, and adapt if necessary.

Organizing Content for Connected Learning

For too many students, the study of mathematics is a daily disconnected experience. The mathematical and statistical processes provide guidance about what students' experiences should look like. The mathematical and statistical content provide opportunities to develop these processes within and across different mathematics concepts. It is important to examine how this content can be organized to support students' experiences to develop the mathematical and statistical processes.

Often educators are provided a list of content standards of disjointed granular actions that students should demonstrate or that should be covered within a course. What should we do with these lists to help students have a successful learning experience that emphasizes connections to other content as well as to real-life situations?

Connecting Concepts

Students see mathematics as a series of disjointed and disconnected topics. Using the Crosscutting Concepts as a key organizing feature in structuring for courses could help students cultivate the relevance and usefulness of the concepts being studied. Teachers and students should be able to identify which Crosscutting Concepts correlate to the topic being studied so that they continually recognize the connections among concepts.

When students recognize they are spending time in ninth grade looking at Making and Interpreting Predictions, for example, and then time in 10th grade addressing Making and Interpreting Predictions, and then additional time in 11th and 12th grade studying Making and Interpreting Predictions, the relevance of the concepts being studied increases as students see increased relationships among topics. Being explicit with students about the Crosscutting Concepts positions them to identify the important aspects of mathematics as well as provides opportunities for students to strengthen their knowledge as they recognize their prior experiences support learning new content.

As you do this work, you may find concepts that do not easily fit into one of the Crosscutting Concepts, and you should consider de-emphasizing them. De-emphasizing these topics allows opportunities for deeper development of the concepts addressing the Crosscutting Concepts and helps students see the relevance and usefulness of what is being studied.

Restructuring Course Content and Experiences

The work of organizing the content into courses is led by state- or district-level educators, often with the participation and input of classroom-level educators, and in some cases, is heavily influenced by curricular materials. As pathways and course structures are determined, attention should shift to designing courses and organizing the content assigned to that course. Often, this work is led by classroom educators.

Ideally this work should be done in teacher teams with others who are teaching the same course in your building or district. Doing so allows for additional perspectives and helps ensure that all students taking the same course experience the same high level of engaged learning as they develop their understanding and use of the same content. This horizontal alignment benefits all students and teachers as the common, shared experiences and understandings can be further developed in future courses.

As a next step, work should be done within the mathematics department to consider how these Crosscutting Concepts can be developed throughout the 4 years of high school. By identifying and organizing content through the five Crosscutting Concepts, vertical awareness and planning can focus on these connections as a means to build more coherence throughout students' high school mathematical experiences. Students will leave high school with a deeper understanding of the mathematics they learned through these connections and will recognize the relevance of what was studied.

Modeling and Statistical Reasoning as Enduring Practices

When learning mathematics, students should develop a skill set that centers on critical thinking, reasoning, and solving meaningful, complex, real-life problems. As educators design courses, it is vital that the classroom experiences help students foster this type of mathematical reasoning and statistical thinking necessary for success. Modeling utilizes these vital skills to apply mathematics and statistics to a relevant context.

Statistical literacy and data science skills and understandings, while of increasing importance, are not currently available to all students. Many students leave high school without developing a base of statistical and data science literacy (Drozda, 2022). Schools and school districts should design mathematics courses and pathways that support all students with the development of modeling and statistical reasoning skills and the ability to apply them to ask and answer questions in the world around them. All mathematics pathways should integrate opportunities for students to explore mathematics in the context of real-life phenomena. The focus should be on creating enduring skills and practices within students that they retain and use beyond high school.

Technology for a New Vision

Technology and the doing of mathematics and the teaching of mathematics are intertwined and at odds from some perspectives. Technology should be used to achieve the goal of students developing enduring mathematical understanding and practices to make sense of and interrogate the world around them.

Technology allows opportunities for students to be engaged in meaningful inquiry related to topics relevant to who they are and what they see as useful and worthwhile.

“Mathematical action technologies”—technologies specifically designed to support the teaching and learning of mathematics and statistics (Dick & Hollebrands, 2011)—can introduce students to mathematics that would have been out of reach without the technology. Students can and should learn to use technologies to apply mathematics and statistics to question and investigate real-life topics, such as the existence of wage gaps by gender, race, education, or occupation (Burrill et al., 2023), to consider different models for representing rising CO_2 levels in the atmosphere, or to determine how they would maximize resources for a projected flood.

Technology plays a central role in the learning process by helping students create robust mental images of concepts through visual dynamic interactive representations. Multiple connected representations, accessed through technology, can deepen understanding by making visible what each representation contributes to the mathematical story (Cullen et al., 2020). For example, when students simultaneously look at symbolic, graphical, numerical, and verbal representations of a context and change one of the parameters, they are immediately able to see and reflect on how the change affected each representation. Technology can also serve to open doors to multiple approaches, enabling students to recognize that there is not just one way to solve a problem.

Conveyance technologies, which display computer or handheld device screens to the whole class, can make students’ thinking and diverse strategies visible, encouraging students to learn from each other and participate in mathematical discussions, giving them opportunities to construct positive mathematical identities of themselves and their peers (Boaler & Staples, 2008; Hufferd-Ackles et al., 2004; Sengupta-Irving, 2014). Encouraging students to share ideas as worthy of exploration provides opportunities for them to be positioned as mathematically competent and allows them to meaningfully participate in the mathematics, personally and socially (Berry & Larson, 2019). The technologies also allow the teachers to monitor students as they work, identifying those who need support and encouraging collaboration among groups approaching the task in similar or different ways. Together, the conveyance and mathematical action technologies can enable teachers to create environments that foster discovery and meaning making in mathematics for all students.

De-emphasizing Content

Interactive dynamic technology can reduce the following rote procedures and instead support the development of students’ thinking and reasoning, allowing time for new content, including statistical reasoning and thinking, analyzing big data sets, and modeling. Much of what has traditionally been taught in mathematics is now easily accessible through technology. As such, we should question the extended focus on traditional algebraic memorization of symbolic manipulations, which are no longer relevant. Too often, mathematical action technology is used in the teaching and learning of mathematics and statistics only to do what had to be done in the past in the absence of the technology (Papert, 2006).

Consider, for example, factoring trinomials, a topic that consumes a large amount of time in beginning and intermediate algebra. Learning how to factor trinomials using a particular method or manipulation is not really a productive goal; understanding factoring polynomials is. Students factored trinomials because it was the only way to introduce them to the general concept of factoring when the only available tools were paper and pencil. Rather, recognizing that factoring a trinomial is not a general competency and thinking about what is important with respect to factoring, perhaps a better list of objectives might include:

- Connect factors to graphs.
- Identify the possible number of linear factors from the degree.
- Identify possible factors for a polynomial and justify the possibilities.
- Factor simple expressions by reasoning about common factors and relationships.

Allowing students access to connections between numeric and graphing tools and computer algebra systems can enable them to think about factors from this broader perspective.

The process of reflecting on what is necessary to understand and effectively use can reduce the emphasis on many of the symbolic procedures students have typically been forced to learn and perform on a test with little or no understanding. This allows the focus to transition from symbolic manipulation to sense making and application for a variety of equations, measures of center and variability, and derivatives and integrals. The technology is a tool to support and engage students in thinking about the concepts, ideas, and relationships in productive ways and to see the usefulness of mathematics.

The role of technology in a new vision of high school mathematics cannot be isolated from the content. Mastery of skills should not be a prerequisite for using technology in any content area; rather, the focus when using technology should be on developing understanding and interpreting the results (NCTM, 2018; Roschelle et al., 2000; Sacristán et al., 2010). As students engage with the mathematics related to each of the five Crosscutting Concepts described in this document, technology should be utilized to open mathematical doors for all students, a driver of what is important to learn and integral to developing an understanding of the content. The goal should be to enable students to recognize which techniques produce a desired outcome, interpret the outcome mathematically, and use the outcome to analyze a situation or solve a problem (NCTM, 2018).

Engagement Through Modeling, Tasks, and Technology

This vision of the high school mathematics classroom, with modeling, tasks, and appropriate use of technology, must be the standard for all student learning, not just a few. In far too many classrooms, students are disengaged, and there is little questioning, discourse, reasoning, or sense making. This is too often the case for students who have been deemed to be unsuccessful or less successful in learning mathematics. Students who are deemed to be successful with mathematics may have a chance to be active participants and effectively utilize technology in modeling and other tasks to develop a deeper understanding of mathematics. Ultimately, all high school mathematics classes must have as a foundation engaging experiences based on quality tasks, learning through modeling experiences that reinforce the utility of mathematics, and the application of the technologies to support students' thinking, reasoning, and application of mathematics.

Changing the foundation of how high school mathematics classes approach their students also has the benefit of significantly impacting how students leave high school, seeing their mathematical abilities and identity. Developing positive student identities must be part of educators' everyday work. The way students view themselves as learners of mathematics greatly influences their willingness to engage and participate (Bishop, 2012; Nasir & Hand 2006). Educators must use teaching practices that focus on mathematics, leverage multiple mathematical competencies, affirm mathematical identities, challenge marginality, and draw on multiple resources of knowledge (Aguirre et al., 2024).

The eight Mathematics Teaching Practices articulated in *Principles to Actions* (NCTM 2014) provide a framework for making connections between instructional practices and the development of student identity. *Catalyzing Change in High School Mathematics* (NCTM 2018) provides the crosswalk (see Appendix C) that shows the relationship between these eight practices and equitable Mathematics Teaching Practices that focus on building and supporting positive student identity and agency. Educators must keep these in mind when making the necessary instructional changes to help students have the experiences needed to see the relevance and usefulness of the concepts being studied.

CHAPTER 5

Continuing the Work

Enacting change is difficult, complex, and extended work. The work of change often begins with casting and setting a vision for the experiences high school students have in learning mathematics; there is also much work that needs to occur with developing an implementation plan, executing the plan, then monitoring, adjusting, and sustaining the success of the changes. Although the work can be initiated by individual educators in their own classrooms, the impact is much greater when these changes translate to sustained and consistent change across all mathematics classes within a school or district and even across a state or province.

PAUSE AND REFLECT: With Your Colleagues: Establishing Goals

- Who is benefiting from the current system? Who is not?
- Do students currently leave high school recognizing the relevance and usefulness of the mathematics they learned?
- Are the skills that students are learning truly beneficial for post–high school success?
- What skills do you hope every child leaves high school with?
- What is our motivation for change?
- What is the motivation for resisting this change?

Collaborating for Change: Critical Conversations

In educational change efforts, collaborating with a wide variety of stakeholders is imperative because it brings together diverse perspectives for critical conversations, fosters ownership and buy-in, generates a wider array of solutions, develops supportive communicators at multiple levels, and creates a foundation for sustainability. Comprehensive collaboration can result in a more well-rounded, effective, and sustainable transformation process that considers the diverse needs and perspectives from the entire educational community, ultimately leading to better outcomes for all students. Depending on the scale of your vision for change, you may consider including the following stakeholders:

- teachers
- parents

- students
- school and district leaders
- community leaders
- business and industry leaders
- higher education faculty
- state-level leaders

To set a vision of student experiences in high school mathematics and guide the work in creating a well-rounded implementation plan, it's crucial to involve all parties in the planning process from the beginning. This inclusive approach allows everyone to understand the current state, develop goals, and contribute their valuable expertise and insights, fostering a more thoughtful and effective strategy. It's important to recognize that the diversity of your stakeholders will result in individuals coming with different perspectives on the current state and what is attainable. We all bring diverse knowledge to the table, enriching the decision-making process and helping anticipate and address potential unseen areas.

Engaging all parties strengthens support for setting the vision and enacting the needed changes. It provides an opportunity to establish clear goals, commitments, and steps for each component of the plan. By involving all from the onset, we can collectively identify challenges and opportunities and brainstorm effective implementation strategies.

Assessing the Current State of the Environment

Before change can happen, it is vital to assess the state of your environment and context to provide an opportunity to analyze the realities of your mathematics instructional program. By reflecting on the status of your current situation, you can ask questions, make comparisons, outline steps, and initiate a plan. The role you serve within the system of high school mathematics likely impacts what aspects of the current state you will be able to assess. For instance, the perspectives of a high school mathematics instructional coach may be different for a high school mathematics teacher. Although both perspectives will have similarities, they will also have some informative differences in the assessments of the current state.

Assessing the Current State as a High School Mathematics Teacher

Engaging teachers in assessing the current state of classroom instruction is crucial to ensure that they are interrogating the effectiveness of their teaching methods for student learning and engagement. There are a number of things high school mathematics teachers can use to assess the current state of student experiences, such as student assessment data, observation data, feedback from students, peer observations, and availability of relevant professional development opportunities. In assessing the current state of instruction, teachers will be focused on what they are currently doing and compare that to what they would be doing to support the common vision.

Assessing the Current State as a School or District Mathematics Instructional Leader

School and district mathematics instructional leaders play an important role in assessing and improving classroom instruction across an entire school and a district with a commitment to equity and inclusion to ensure that all students have access to quality instruction and are experiencing success. Leaders can collect and analyze data on student performance, conduct classroom observations, and seek feedback from teachers, students, parents, and the community. They also assess the alignment of the curriculum to the mission and vision of the program. By evaluating curriculum materials, professional development programs, and the effectiveness of instructional strategies, school and district mathematics leaders gain a more comprehensive view of the district's educational landscape, including strengths and areas of need. This assessment can support strategic planning, resource allocation, and continuous improvement efforts, enabling leaders to elevate the overall quality of education and support the growth and development of both students and teachers.

Assessing the Current State as a School or District Administrator

Engaging several educational partners, including nonmathematics leaders, in the process of assessing the current state and setting a vision for change is vital to maximize buy-in and support. School leaders have a responsibility to ensure that mathematics instruction leads to positive experiences for every student and is structured to increase, not restrict, the student's opportunities. How is the vision for education as a whole reflected in the mathematics program and for all students? How can the goals of the mathematics program be used to communicate to the community and all stakeholders the *why* of the change?

As leaders assess a mathematics program, they should determine ways to assess student performance data, graduation data, and postsecondary plans data to determine if the mathematics program is ensuring student success. This assessment should examine the impact of current pedagogical approaches, mathematical content, structures, and professional support of a variety of educational partners.

Assessing the Current State as a State Mathematics Instructional Leader

State mathematics instructional leaders can employ a range of strategies to assess the current state of mathematics learning and classroom instruction across their educational systems. They rely on comprehensive data analysis, including standardized test scores and course taking and completion data, to gauge student performance and identify targeted and statewide programmatic needs. State leaders also engage with school and district administrators, mathematics leaders and teachers, students, and other stakeholders through surveys and other feedback mechanisms to gain perspectives. Classroom observations can provide insights into instructional practices, allowing them to assess teaching quality and the alignment of instruction with state standards. They may also conduct curriculum reviews by examining materials and resources for alignment and quality. This multifaceted approach informs policymaking, funding allocation, and the development of targeted support initiatives, ultimately ensuring the enhancement of classroom instruction statewide and increasing student achievement and success.

PAUSE AND REFLECT: Cultivating Support and Considering Challenges

- What concerns or interests of the school district, state or province, and community support a reimagined vision for mathematics pathways?
- What beliefs are held by leaders and community members regarding high school mathematics courses and instruction? How do these beliefs align with the reimagined vision for high school mathematics?
- What practices and beliefs are held by teachers or teacher teams about mathematics instruction? How do these practices and beliefs align with a reimagined vision for high school mathematics?
- Who currently determines what mathematics courses students take in your school(s)? How will these individuals be engaged in the work?
- What constraints could you have in terms of policies, practices, and/or funding?
- What opportunities and challenges exist to support collaboration between mathematics teachers in the building, across the district and/or state or province?
- What other challenges do you anticipate as you begin to develop and enact a reimagined vision for mathematics pathways?

Cultivate a Common Vision

It is critical for everyone involved in the work of reimagining and revitalizing high school mathematics to establish a common and clear vision that will inform the necessary steps to initiate and sustain change. This task is often the role of administrators and mathematics leaders in their schools and districts (NCSM, 2019). When you establish a common vision for your high school mathematics program, consider the profound impact this work can have on shaping students' mathematical identities and preparing them for the future. Students must be active participants in their mathematical journey and be encouraged to ask questions, discover connections, and apply their knowledge to real-world problems. Be sure to weave inclusivity and diversity of thought into this vision to ensure that all students, regardless of their background, feel valued and empowered.

To ensure high school students have a positive experience with mathematics, believe in themselves, and leave with opportunities for future success, here again are the three core actions that teachers and individuals at all levels need to implement.

1. **Relevance becomes a defining characteristic of mathematics classrooms and learning** through mathematical and statistical modeling and the use of contextual and interesting tasks.
2. **Reimagine the content by organizing around Crosscutting Concepts and creating interest-driven pathways** to deepen students' understanding of mathematics and statistics, highlight connections between the mathematical and statistical concepts, and allow students to recognize the overall utility of the mathematics they are learning.
3. **Revitalize the student experience by using mathematical and statistical processes** to ensure student engagement, active sense making of mathematics and statistics, and use of mathematics and statistics to question, understand, or critique the world around them.

Merely addressing one or two core actions will not be sufficient to transform high school mathematics and develop students who see the relevance and utility of the mathematics they are learning. These focused changes to structures, content, and pedagogy are designed for students to leave high school mathematics, seeing themselves as capable of continuing to learn and apply mathematics in their lives.

Taking the Lead for Teachers

To inspire to continue the necessary work of reimagining systems and structures for engaging high school students in relevant and useful math learning and fostering their self-belief, *teachers* can initiate the following steps:

- Examine how the collective mathematical and statistical content and processes develop throughout students' mathematical learning experiences in high school.
- Reimagine the sequence and approach to topics to highlight the five Crosscutting Concepts and reduce time on content with limited utility and engagement.
- Using the Crosscutting Concepts, determine which understanding and applications should be emphasized, including through mathematical and statistical modeling, as well as what can be de-emphasized.
- Consider where and how technology could be used to more effectively allow students more access to and understanding and application of mathematics and statistics.
- Define teaching and learning structures that engage students in making sense of the content they are learning and its application, including connections to other mathematics and statistics concepts and applications.

- Implement the eight Mathematics Teaching Practices consistently and equitably so that all students leave high school with rich mathematical and statistical experiences and enduring practices.

Taking the Lead for Schools and Districts

To inspire to continue the necessary work of reimagining systems and structures for engaging high school students in relevant and useful math learning and fostering their self-belief, *schools and districts* can initiate the following steps:

- Engage community members to understand and support needed changes in high school mathematics to create positive, enduring mathematical and statistical practices for each student and ensure those changes collectively work to increase opportunities for each student.
- Provide time and support for teachers, mathematics leaders, and school and district administrators to collaborate to create and organize interest-driven pathways that offer students opportunities to develop a deep understanding of mathematics and statistics.
- Provide support for teachers as they learn and implement new teaching practices and integrate mathematical and statistical modeling into their instruction.
- Use the Crosscutting Concepts to help determine which content should be emphasized as well as what can be de-emphasized.
- Support teachers with professional learning opportunities to collaborate on instructional strategies to develop mathematical and statistical processes with students.

Taking the Lead for Policymakers

To inspire to continue the necessary work of reimagining systems and structures for engaging high school students in relevant and useful math learning and fostering their self-belief, *policymakers* can initiate the following steps:

- Engage a wide range of stakeholders to develop support for needed changes in high school mathematics.
- Develop policies that support meaningful interest-driven pathways that support students' development of the Crosscutting Concepts as they learn and apply mathematics.
- Develop processes that are aligned with and emphasize the mathematical and statistical processes and the Crosscutting Concepts to evaluate systemic progress and provide support for long-term change.
- Focus support for creating multiple pathways with high expectations, quality instruction, and an emphasis on increasing students' opportunities.

Taking the Lead for Postsecondary Educators

To inspire to continue the necessary work of reimagining systems and structures for engaging high school students in relevant and useful math learning and fostering their self-belief, *postsecondary educators* can initiate the following steps:

- Work with in-service and preservice teachers to implement research-informed equitable instructional practices to help students build their mathematical understandings and see the relevance and utility of mathematics.
- Collaborate with school and district educators to develop interest-driven pathways to prepare students for success after high school graduation.
- Collaborate with school-, district-, and state-based mathematics educators to develop structures to reinforce and support educators during their transition and induction into secondary teaching.

Taking the Lead for Professional Organizations

To inspire to continue the necessary work of reimagining systems and structures for engaging high school students in relevant and useful math learning and fostering their self-belief, *professional organizations* can initiate the following steps:

- Share examples of districts and states that have engaged in this work, and provide resources and support for leaders undertaking these efforts.
- Provide additional resources to help educators fully integrate the mathematical and statistical processes into their instruction.
- Provide professional learning opportunities to provide additional examples and resources clarifying and illustrating the key major recommendations.

In This Together

This work is challenging and will take significant time, but it is important to provide mathematical experiences that help students see the relevance and utility of what they are learning. As educators have done in the past, we must recognize that this work does not need to be done in isolation. We can learn from and with each other and build on others' work to transform student experiences in high school mathematics through interest-driven pathways.

APPENDIX A

Mathematical and Statistical Modeling Topics

Following is a list of examples of topics for mathematical and statistical modeling that high school students may find interesting and engaging and that allow students to ask and answer questions relevant to their surroundings and interests.

1. **Health and Nutrition**
   - **Modeling the Relationship Between Sleep and Academic Performance:** Use statistical analysis to explore the correlation between the amount of sleep students get and their grades or test scores.
   - **Analyzing the Impact of Diet on Physical Fitness:** Collect data on dietary intake and physical fitness tests to model how changes in diet affect fitness levels.
   - **Studying the Spread of Seasonal Flu in Schools:** Model the spread of flu in a school community based on vaccination rates, attendance records, and reported cases to understand how diseases spread and are controlled.
2. **Environmental Impact Studies**
   - **Modeling the Reduction in Pollution With Electric Vehicles:** Estimate the decrease in carbon emissions in a community if a certain percentage of residents switched to electric vehicles.
   - **Predicting the Impact of Deforestation on Local Climate:** Use data on local vegetation cover to model how changes in forest area could affect temperature and precipitation patterns.
   - **Analyzing the Effectiveness of Recycling Programs:** Collect and analyze data on waste production and recycling rates to model the impact of recycling programs on reducing landfill use.
3. **Sports Performance Analysis**
   - **Comparing Pre- and Posttraining Performance:** Collect data on athletes before and after a specific training regimen to model improvements in speed, strength, or endurance.
   - **Predicting Game Outcomes Based on Team Statistics:** Use regression analysis to predict the outcomes of games based on team statistics such as average points per game, defensive efficiency, or turnovers.
   - **Analyzing the Effect of Home Court Advantage:** Model the impact of playing at home versus away on team performance throughout a season using a variety of statistical measures.

4. **Financial Literacy**
    - **Modeling Savings Growth Over Time:** Use compound interest formulas to model how savings grow over time under different interest rates and saving strategies.
    - **Analyzing the Cost-Effectiveness of College Choices:** Compare the long-term financial impact of attending different colleges, taking into account tuition, potential scholarships, and expected starting salaries.
    - **Projecting Future Expenses and Savings:** Create a model to project future expenses (like college, travel, or buying a car) and how current savings behavior can meet those goals.
5. **Social Media Trends**
    - **Modeling the Spread of a Hashtag:** Analyze how quickly a hashtag spreads on social media platforms and identify key factors in its proliferation.
    - **Predicting the Popularity of New Social Media Features:** Use historical data on feature usage to model the potential popularity of newly released social media platform features.
    - **Analyzing Influencer Impact on Product Sales:** Model the relationship between social media influencer promotions and subsequent product sales, accounting for variables like follower count and engagement rate.
6. **Traffic Flow and Public Transportation**
    - **Modeling Traffic Patterns Around School Events:** Analyze how events like sports games or concerts affect local traffic patterns and propose solutions to minimize congestion.
    - **Predicting Public Transportation Usage Based on Weather:** Use weather data to model changes in public transportation usage and suggest adjustments to schedules or capacity.
    - **Analyzing the Efficiency of Different Commute Routes:** Collect data on travel times for various routes and modes of transportation to model the most efficient commuting strategies.
7. **Educational Outcomes**
    - **Modeling the Impact of Study Time on Test Scores:** Use linear regression to explore how different amounts of study time are correlated with performance on standardized tests.
    - **Analyzing the Role of Extracurricular Activities in College Admissions:** Collect data on extracurricular participation and college admission outcomes to model the impact of various activities.
    - **Predicting Grade Trends Based on Classroom Engagement:** Model how factors like attendance, participation, and homework completion correlate with changes in grades over time.

8. **Weather and Climate Analysis**
   - **Predicting Seasonal Weather Patterns for Agricultural Planning:** Use historical weather data to model seasonal patterns and assist in planning planting or harvesting activities.
   - **Modeling the Impact of Urbanization on Local Weather:** Analyze how changes in land use and urban development affect local temperature, runoff and flooding, and heat retention.
   - **Estimating the Effects of Global Warming on Local Species:** Model predictions for temperature and habitat changes to estimate the potential impact on local wildlife populations and biodiversity.

Following is a list of examples of topics for statistical reasoning that high school students may find interesting and engaging and allow students to ask and answer questions relevant to their surroundings and interests.

1. **Representation, Services, and Opportunities Analysis**
   - **Food Deserts:** Students collect data on the number of places to acquire healthy food and the distance needed to travel for various communities and populations and statistically analyze consistencies and differences.
   - **Access to Services:** Students can choose a service or opportunity and compare access between various communities. Topics could include health care, outdoor spaces, daycare, internet access, or other topics of their interest.
   - **Electoral District Boundaries:** Students collect data on representation of various demographic groups within in a state or county and the representation of those populations within electoral district boundaries.
2. **Social Network Influence Analysis**
   - **Comparing Likes and Shares:** Students can collect data on the number of likes and shares for different types of posts on social media platforms to statistically analyze which content types are most engaging.
   - **Influence of Posting Time on Engagement:** By gathering data on the time of day posts are made and the corresponding engagement levels (likes, comments, shares), students can use statistical methods to determine the optimal times for posting.
   - **Effect of Hashtags on Visibility:** Analyze how the use of different hashtags affects the reach and engagement of posts, using statistical tests to compare the effectiveness of various hashtag strategies.
3. **Dietary Impacts on Health Evaluation**
   - **Correlation Between Sugar Intake and Energy Levels:** Students can track their daily sugar consumption and rate their energy levels, using correlation analysis to explore the relationship between sugar intake and perceived energy.

- **Statistical Analysis of Weight Change and Diet Type:** By recording weight changes over time in relation to different diet types (e.g., high protein vs. vegetarian), students can use statistical methods to analyze the effectiveness of these diets.
- **Impact of Breakfast on Academic Performance:** Collect data on breakfast habits and academic performance indicators (test scores, grades) to statistically examine the effect of eating breakfast on school performance.

4. **School Sports Teams Performance Evaluation**
   - **Player Performance Before and After Coaching Changes:** Analyze statistical differences in player performance metrics before and after a coaching change to evaluate the impact of different coaching strategies.
   - **Team Performance in Various Weather Conditions:** Collect data on team performance metrics (e.g., scores, wins) under different weather conditions and use statistical analysis to determine if performance significantly varies with weather.
   - **Statistical Comparison Between During-School and After-School Practice Sessions:** Compare the effectiveness of during-school versus after-school practice sessions on team performance by analyzing game outcomes statistically.
5. **Understanding Environmental Actions**
   - **Recycling Habits and Waste Reduction:** Survey students on their recycling habits and collect data on waste reduction in the school to use statistical methods for analyzing the impact of recycling education on waste output.
   - **Statistical Analysis of Water Usage Before and After Conservation Measures:** Measure water usage in the school or community before and after implementing water conservation measures, using statistical tests to evaluate the effectiveness of these measures.
   - **Effectiveness of Different Types of Reusable Containers:** Gather data on the usage rates of different types of reusable containers (e.g., water bottles, lunch boxes) and their impact on reducing single-use plastics, applying statistical analysis to compare effectiveness.
6. **Assessment of Financial Literacy Programs**
   - **Pre- and Postprogram Financial Knowledge:** Use statistical tests to compare students' scores on financial literacy tests before and after participating in a financial education program to assess its effectiveness.
   - **Correlation Between Financial Literacy and Saving Habits:** Collect data on students' saving habits and their scores on financial literacy assessments to statistically explore the relationship between financial knowledge and behavior.
   - **Statistical Analysis of Allowance Management Strategies:** Survey students on their allowance or part-time job earnings management strategies and analyze data to identify statistically significant patterns in saving or spending behaviors.

APPENDIX B

Linking Essential Concepts to Crosscutting Concepts

Linking Essential Concepts From *Catalyzing Change in High School Mathematics* (NCTM, 2018) to Crosscutting Concepts	
Essential Concepts in Number	**Crosscutting Concepts**
• Together, irrational numbers and rational numbers complete the real number system, representing all points on the number line.	Functional and Structural Thinking
• Quantitative reasoning includes, and mathematical modeling requires, attention to units of measurement.	Comparison, Difference, and Equivalence
Essential Concepts in Algebra and Functions	**Crosscutting Concepts**
Algebra	
• Expressions can be rewritten in equivalent forms by using algebraic properties, including properties of addition, multiplication, and exponentiation, to make different characteristics or features visible.	Comparison, Difference, and Equivalence
• Finding solutions to an equation, inequality, or system of equations or inequalities requires the checking of candidate solutions, whether generated analytically or graphically, to ensure that solutions are found and that those found are not extraneous.	Comparison, Difference, and Equivalence
• The structure of an equation or inequality (including, but not limited to, one-variable linear and quadratic equations, inequalities, and systems of linear equations in two variables) can be purposefully analyzed (with and without technology) to determine an efficient strategy to find a solution, if one exists, and then to justify the solution.	Functional and Structural Thinking
• Expressions, equations, and inequalities can be used to analyze and make predictions, both within mathematics and as mathematics is applied in different contexts—in particular, contexts that arise in relation to linear, quadratic, and exponential situations.	Making and Interpreting Predictions

(continued)

Linking Essential Concepts From *Catalyzing Change in High School Mathematics* (NCTM, 2018) to Crosscutting Concepts *(continued)*

Essential Concepts in Algebra and Functions	Crosscutting Concepts
Connecting Algebra to Functions	
• Functions shift the emphasis from a point-by-point relationship between two variables (input/output) to considering an entire set of ordered pairs (where each first element is paired with exactly one second element) as an entity with its own features and characteristics.	Functional and Structural Thinking
• Graphs can be used to obtain exact or approximate solutions of equations, inequalities, and systems of equations and inequalities—including systems of linear equations in two variables and systems of linear and quadratic equations (given or obtained by using technology).	Patterns and Generalization
Functions	
• Functions can be described by using a variety of representations: mapping diagrams, function notation (e.g., $f(x) = x^2$), recursive definitions, tables, and graphs.	Functional and Structural Thinking
• Functions that are members of the same family have distinguishing attributes (structure) common to all functions within that family.	Functional and Structural Thinking
• Functions can be represented graphically, and key features of the graphs, including zeros, intercepts, and, when relevant, rate of change and maximum/minimum values, can be associated with and interpreted in terms of the equivalent symbolic representation.	Comparison, Difference, and Equivalence
• Functions model a wide variety of real situations and can help students understand the processes of making and changing assumptions, assigning variables, and finding solutions to contextual problems.	Variability and Change
Essential Concepts in Statistics and Probability	**Crosscutting Concepts**
Quantitative Literacy	
• Mathematical and statistical reasoning about data can be used to evaluate conclusions and assess risks.	Making and Interpreting Predictions
• Making and defending informed data-based decisions is a characteristic of a quantitatively literate person.	Comparison, Difference, and Equivalence

Linking Essential Concepts From *Catalyzing Change in High School Mathematics* (NCTM, 2018) to Crosscutting Concepts *(continued)*

Essential Concepts in Statistics and Probability	Crosscutting Concepts
Visualizing and Summarizing Data	
• Data arise from a context and come in two types: quantitative (continuous or discrete) and categorical. Technology can be used to "clean" and organize data, including very large data sets, into a useful and manageable structure—a first step in any analysis of data.	Patterns and Generalization
• Distributions of quantitative data (continuous or discrete) in one variable should be described in the context of the data with respect to what is typical (the shape, with appropriate measures of center and variability, including standard deviation) and what is not (outliers), and these characteristics can be used to compare two or more subgroups with respect to a variable.	Variability and Change
• The association between two categorical variables is typically represented by using two-way tables and segmented bar graphs.	Patterns and Generalization
• Scatterplots, including plots over time, can reveal patterns, trends, clusters, and gaps that are useful in analyzing the association between two contextual variables.	Patterns and Generalization
• Analyzing the association between two quantitative variables should involve statistical procedures, such as examining (with technology) the sum of squared deviations in fitting a linear model, analyzing residuals for patterns, generating a least-squares regression line and finding a correlation coefficient, and differentiating between correlation and causation.	Making and Interpreting Predictions
• Data analysis techniques can be used to develop models of contextual situations and to generate and evaluate possible solutions to real problems involving those contexts.	Variability and Change
Statistical Inference	
• Study designs are of three main types: sample survey, experiment, and observational study.	Comparison, Difference, and Equivalence
• The role of randomization is different in randomly selecting samples and in randomly assigning subjects to experimental treatment groups.	Functional and Structural Thinking

(continued)

Linking Essential Concepts From *Catalyzing Change in High School Mathematics* (NCTM, 2018) to Crosscutting Concepts *(continued)*

Essential Concepts in Statistics and Probability	Crosscutting Concepts
Statistical Inference	
• The scope and validity of statistical inferences are dependent on the role of randomization in the study design.	Functional and Structural Thinking
• Bias, such as sampling, response, or nonresponse bias, may occur in surveys, yielding results that are not representative of the population of interest.	Variability and Change
• The larger the sample size, the less the expected variability in the sampling distribution of a sample statistic.	Variability and Change
• The sampling distribution of a sample statistic formed from repeated samples for a given sample size drawn from a population can be used to identify typical behavior for that statistic. Examining several such sampling distributions leads to estimating a set of plausible values for the population parameter, using the margin of error as a measure that describes the sampling variability.	Patterns and Generalization
• Simulation of sampling distributions by hand or with technology can be used to determine whether a statistic (or statistical difference) is significant in a statistical sense or whether it is surprising or unlikely to happen under the assumption that outcomes are occurring by random chance.	Comparison, Difference, and Equivalence
Probability	
• Two events are independent if the occurrence of one event does not affect the probability of the other event. Determining whether two events are independent can be used for finding and understanding probabilities.	Functional and Structural Thinking
• Conditional probabilities—that is, those probabilities that are "conditioned" by some known information—can be computed from data organized in contingency tables. Conditions or assumptions may affect the computation of a probability.	Functional and Structural Thinking
Measurement	
• Areas and volumes of figures can be computed by determining how the figure might be obtained from simpler figures by dissection and recombining.	Patterns and Generalization

Linking Essential Concepts From *Catalyzing Change in High School Mathematics* (NCTM, 2018) to Crosscutting Concepts *(continued)*	
Essential Concepts in Statistics and Probability	**Crosscutting Concepts**
Measurement	
▪ Constructing approximations of measurements with different tools, including technology, can support an understanding of measurement.	Comparison, Difference, and Equivalence
▪ When an object is the image of a known object under a similarity transformation, a length, area, or volume on the image can be computed by using proportional relationships.	Patterns and Generalization
Transformations	
▪ Applying geometric transformations to figures provides opportunities for describing the attributes of the figures preserved by the transformation and for describing symmetries by examining when a figure can be mapped onto itself.	Comparison, Difference, and Equivalence
▪ Showing that two figures are congruent involves showing that there is a rigid motion (translation, rotation, reflection, or glide reflection) or, equivalently, a sequence of rigid motions that maps one figure to the other.	Comparison, Difference, and Equivalence
▪ Showing that two figures are similar involves finding a similarity transformation (dilation or composite of a dilation with a rigid motion) or, equivalently, a sequence of similarity transformations that maps one figure onto the other.	Comparison, Difference, and Equivalence
▪ Transformations in geometry serve as a connection with algebra, both through the concept of functions and through the analysis of graphs of functions as geometric figures.	Functional and Structural Thinking
Geometric Arguments, Reasoning, and Proof	
▪ Proof is the means by which we demonstrate whether a statement is true or false mathematically, and proofs can be communicated in a variety of ways (e.g., two-column, paragraph).	Comparison, Difference, and Equivalence
▪ Using technology to construct and explore figures with constraints provides an opportunity to explore the independence and dependence of assumptions and conjectures.	Patterns and Generalization

(continued)

Linking Essential Concepts From *Catalyzing Change in High School Mathematics* (NCTM, 2018) to Crosscutting Concepts *(continued)*

Essential Concepts in Statistics and Probability	Crosscutting Concepts
Geometric Arguments, Reasoning, and Proof	
• Proofs of theorems can sometimes be made with transformations, coordinates, or algebra; all approaches can be useful, and in some cases one may provide a more accessible or understandable argument than another.	Functional and Structural Thinking
Solving Applied Problems and Modeling in Geometry	
• Recognizing congruence, similarity, symmetry, measurement opportunities, and other geometric ideas, including right triangle trigonometry in real-world contexts, provides a means of building understanding of these concepts and is a powerful tool for solving problems related to the physical world in which we live.	Comparison, Difference, and Equivalence
• Experiencing the mathematical modeling cycle in problems involving geometric concepts, from the simplification of the real problem through the solving of the simplified problem, the interpretation of its solution, and the checking of the solution's feasibility, introduces geometric techniques, tools, and points of view that are valuable to problem-solving.	Patterns and Generalization

APPENDIX C

Math Teaching Practices Supporting Equitable Mathematics Instruction

Mathematics Teaching Practices: Supporting Equitable Mathematics Teaching	
Mathematics Teaching Practices	**Equitable Teaching**
Establish mathematics goals to focus learning. Effective teaching of mathematics establishes clear goals for the mathematics that students are learning, situates goals within learning progressions, and uses the goals to guide instructional decisions.	• Establish learning progressions that build students' mathematical understanding, increase their confidence, and support their mathematical identities as doers of mathematics. • Establish high expectations to ensure that each and every student has the opportunity to meet the mathematical goals. • Establish classroom norms for participation that position each and every student as a competent mathematics thinker. • Establish classroom environments that promote learning mathematics as just, equitable, and inclusive.
Implement tasks that promote reasoning and problem-solving. Effective teaching of mathematics engages students in solving and discussing tasks that promote mathematical reasoning and problem-solving and allow multiple entry points and varied solution strategies.	• Engage students in tasks that provide multiple pathways for success and that require reasoning, problem-solving, and modeling, thus enhancing each student's mathematical identity and sense of agency. • Engage students in tasks that are culturally relevant. • Engage students in tasks that allow them to draw on their funds of knowledge (i.e., the resources that students bring to the classroom, including their home, cultural, and language experiences).
Use and connect mathematical representations. Effective teaching of mathematics engages students in making connections among mathematical representations to deepen understanding of mathematics concepts and procedures and to use as tools for problem-solving.	• Use multiple representations so that students draw on multiple resources of knowledge to position them as competent. • Use multiple representations to draw on knowledge and experiences related to the resources that students bring to mathematics (culture, contexts, and experiences). • Use multiple representations to promote the creation and discussion of unique mathematical representations to position students as mathematically competent.

(continued)

Mathematics Teaching Practices: Supporting Equitable Mathematics Teaching *(continued)*	
Mathematics Teaching Practices	**Equitable Teaching**
Facilitate meaningful mathematical discourse. Effective teaching of mathematics facilitates discourse among students to build shared understanding of mathematical ideas by analyzing and comparing student approaches and arguments.	• Use discourse to elicit students' ideas and strategies and create space for students to interact with peers to value multiple contributions and diminish hierarchal status among students (i.e., perceptions of differences in smartness and ability to participate). • Use discourse to attend to ways in which students position one another as capable or not capable of doing mathematics. • Make discourse an expected and natural part of mathematical thinking and reasoning, providing students with the space and confidence to ask questions that enhance their own mathematical learning. • Use discourse as a means to disrupt structures and language that marginalize students.
Pose purposeful questions. Effective teaching of mathematics uses purposeful questions to assess and advance students' reasoning and sense making about important mathematical ideas and relationships.	• Pose purposeful questions and then listen to and understand students' thinking to signal to students that their thinking is valued and makes sense. • Pose purposeful questions to assign competence to students. Verbally mark students' ideas as interesting or identify an important aspect of students' strategies to position them as competent. • Be mindful of the fact that the questions that a teacher asks a student and how the teacher follows up on the student's response can support the student's development of a positive mathematical identity and sense of agency as a thinker and doer of mathematics.
Build procedural fluency from conceptual understanding. Effective teaching of mathematics builds fluency with procedures on a foundation of conceptual understanding so that students, over time, become skillful in using procedures flexibly as they solve contextual and mathematical problems.	• Connect conceptual understanding with procedural fluency to help students make sense of the mathematics and develop a positive disposition toward mathematics. • Connect conceptual understanding with procedural fluency to reduce mathematical anxiety and position students as mathematical knowers and doers. • Connect conceptual understanding with procedural fluency to provide students with a wider range of options for entering a task and building mathematical meaning.
Support productive struggle in learning mathematics. Effective teaching of mathematics consistently provides students, individually and collectively, with opportunities and supports to engage in productive struggle as they grapple with mathematical ideas and relationships.	• Allow time for students to engage with mathematical ideas to support perseverance and identity development. • Hold high expectations, while offering just enough support and scaffolding to facilitate student progress on challenging work, to communicate caring and confidence in students.

Mathematics Teaching Practices: Supporting Equitable Mathematics Teaching *(continued)*	
Mathematics Teaching Practices	**Equitable Teaching**
Elicit and use evidence of student thinking. Effective teaching of mathematics uses evidence of student thinking to assess progress toward mathematical understanding and to adjust instruction continually in ways that support and extend learning.	• Elicit student thinking and make use of it during a lesson to send positive messages about students' mathematical identities. • Make student thinking public, and then choose to elevate a student to a more prominent position in the discussion by identifying his or her idea as worth exploring, to cultivate a positive mathematical identity. • Promote a classroom culture in which mistakes and errors are viewed as important reasoning opportunities, to encourage a wider range of students to engage in mathematical discussions with their peers and the teacher.

Source. National Council of Teachers of Mathematics. (2018). *Catalyzing change in high school mathematics: Initiating critical conversations.*

References

Abassian, A., Safi, F., Bush, S., & Bostic, J. (2020). Five different perspectives on mathematical modeling in mathematics education. *Investigations in Mathematics Learning, 12*(1), 53–65. https://doi:10.1080/19477503.2019.1595360

ACT. (2019). *The condition of college & career readiness 2019.* https://www.act.org/content/dam/act/unsecured/documents/National-CCCR-2019.pdf

Aguirre, J. M., Mayfield-Ingram, K., & Martin, D. B. (2024). *The impact of identity in K–12 mathematics: Rethinking equity-based practices*, Expanded Edition. National Council of Teachers of Mathematics.

Amrein, A., & Berliner, D. (2003). The effects of high-stakes testing on student motivation and learning. *Educational Leadership*, 60.

Anderson, V., & Burdman, P. (2022). *A new calculus for college admissions: How policy, practice, and perceptions of high school math education limit equitable access to college.* Just Equations. https://justequations.org/resource/a-new-calculus-for-college-admissions-how-policy-practice-and-perceptions-of-high-school-math-education-limit-equitable-access-to-college

Andre, M., & Zsolt, L. (2019). Technology changing statistics education: Defining possibilities, opportunities and obligations. *Electronic Journal of Mathematics and Technology*, *13*, 253–264.

Bailey, N., & McCulloch, A. (2023). Describing critical statistical literacy habits of mind. *Journal of Mathematical Behavior*, *70*, 101063. https://doi:10.1016/j.jmathb.2023.101063

Banilower, E. R., Smith, P. S., Malzahn, K. A., Plumley, C. L., Gordon, E. M., & Hayes, M. L. (2018). *Report of the 2018 NSSME+*. Horizon Research, Inc.

Bargagliotti, A., Franklin, C., Arnold, P., Gould, R., Johnson, S., Perez, L., & Spangler, D. (2020). *Pre-K–12 Guidelines for Assessment and Instruction in Statistics Education (GAISE) report II.* American Statistical Association and National Council of Teachers of Mathematics.

Bengmark, S., Thunberg, H., & Winberg, M. (2017). Success-factors in transition to university mathematics. *International Journal of Mathematical Education in Science and Technology*, *48*, 1–14. https://doi:10.1080/0020739X.2017.1310311

Berry, R., III, & Larson, M. (2019). The need to catalyze change in high school mathematics. *Phi Delta Kappan*, *100*(6), 39–44.

Berry, R. Q., III, Conway, B. M., IV, Lawler, B. R., & Staley, J. W. (2020). *High school mathematics lessons to explore, understand, and respond to social injustice.* Sage Publications.

Bishop, J. P. (2012, January). She's always been the smart one. I've always been the dumb one: Identities in the mathematics classroom. *Journal for Research in Mathematics Education*, *43*(1), 34–74.

Boaler, J. (2016). *Mathematical mindsets: Unleashing student's potential through creative math, inspiring messages and innovative teaching.* Jossey-Bass.

Boaler, J., & Staples, M. (2008). Creating mathematical futures through an equitable teaching approach: The case of Railside School. *Teachers College Record*, *110*(3), 608–645. https://doi.org/10.1177/016146810811000302

Boston, M, Dillon, F., Smith, M. S., & Miller, S. (2017). *Taking action: Implementing effective mathematics teaching practices, grades 9–12.* National Council of Teachers of Mathematics.

Burdman, P. (2018, November). *The mathematics of opportunity: Rethinking the role of math in educational equity*. https://justequations.org/resource/the-mathematics-of-opportunity-rethinking-the-role-of-math-in-educational-equity

Burrill, G., Funderburk, J., Byer, B., & Gorsuch, R. (2023). Using technology to explore the wage gap. *Mathematics Teacher: Learning and Teaching Mathematics PreK–12, 116*(5), 378–386. https://doi.org/10.5951/MTLT.2022.0348

Center for Effective Philanthropy, Inc. (2023). *Making sense of learning math: Insights from the student experience*. https://cep.org/blog/what-we-learn-when-we-listen-student-feedback-and-foundation-strategy/

Cirillo, M., & Pelesko, J. (2022). *Unlocking the mystery: Mathematical modeling in secondary classrooms.* Math Solutions Publications.

Cullen, C., Hertel, J., & Nickels, M. (2020). The roles of technology in mathematics education. *Educational Forum, 84*(2):166–178. https://doi:10.1080/00131725.2020.1698683

Cuoco, A., Goldenberg, E. P., & Mark, J. (1996). Habits of mind: An organizing principle for mathematics curricula. *Journal of Mathematical Behavior, 15*(4), 375–402. ISSN 0732-3123, https://doi.org/10.1016/S0732-3123(96)90023-1

Cuoco, A., Goldenberg, E. P., & Mark, J. (2010). Contemporary curriculum issues: Organizing a curriculum around mathematical habits of mind. *Mathematics Teacher, 103*(9), 682–688.

Czocher, J., Melhuish, K., & Kandasamy, S. (2019). Building mathematics self-efficacy of STEM undergraduates through mathematical modelling. *International Journal of Mathematical Education in Science and Technology, 51*, 1–28. 10.1080/0020739X.2019.1634223

Dick, T., & Hollebrands, K. (2011). *Focus in high school mathematics: Technology to support reasoning and sense making*. National Council of Teachers of Mathematics.

Drozda, Z. (2022). *Data science is vital to student success. So why are outcomes going down?* Data Science 4 Everyone. https://www.datascience4everyone.org/_files/ugd/d2c47c_067273b2bef041c7ae08cab6c7a3be8c.pdf

EdSource. (2012, February). *Passing when it counts*. https://edsource.org/wp-content/publications/pub12-Math2012Final.pdf

Ferreira, C. (2006). *Gene expression programming: Mathematical modeling by an artificial intelligence* (Vol. 21). Springer.

Garfunkel, S., & Montgomery, M. (Eds.). (2019). *GAIMME: Guidelines for Assessment and Instruction in Mathematical Modeling Education* (2nd ed.). Consortium for Mathematics and its Applications (COMAP) and Society for Industrial and Applied Mathematics (SIAM).

Gordon, S. P. (2008). "What's wrong with college algebra?" *PRIMUS, 18*(6), 516–541. https://doi:10.1080/10511970701598752

Grouws, D. A., Tarr, J. E., Chavez, O., Seats, R., Soria, V. M., & Taylan, R. D. (2013). Curriculum and implementation effects on high school students' mathematics learning from curricula representing subject-specific and integrated content organization. *Journal of Research of Mathematics Education, 44*(2), 416–463.

Hartzell, G. E., Priest, D. N., & Switzer, W. G. (2023). Modeling of toxicological effects of fire gases: II. Mathematical modeling of intoxication of rats by carbon monoxide and hydrogen cyanide. In *Advances in Combustion Toxicology* (Vol. I) (pp. 252–265). CRC Press.

Herbel-Eisenmann, B., Sinclair, N., Chval, K., Clements, D., Civil, M., Pape, S., Stephan, M., Wanko, J., & Wilkerson, T. (2016). Positioning mathematics education researchers to influence storylines. *Journal for Research in Mathematics Education*, *47*(102). 10.5951/jresematheduc.47.2.0102

Hirsch, C., & McDufffie, A. R. (2016). *Annual perspectives in mathematics education 2016: Mathematical modeling*. National Council of Teachers of Mathematics.

Hufferd-Ackles, K., Fuson, K., & Sherin, M. (2004). Describing levels and components of a math-talk learning community. *Journal for Research in Mathematics Education*, *35*(2), 81–116. https://doi.org/10.2307/30034933

Irwin, V., Wang, K., Tezil, T., Zhang, J., Filbey, A., Jung, J., Bullock Mann, F., Dilig, R., & Parker, S. (2023). *Report on the condition of education 2023* (NCES 2023–144). U.S. Department of Education. National Center for Education Statistics. https://nces.ed.gov/pubsearch/pubsinfo.asp?pubid=2023144

Jimenez, L., Sargrad, S., Morales, J., & Thompson, M. (2016). *Remedial education: The cost of catching up.* Center for American Progress. https://www.americanprogress.org/issues/education-k-12/reports/2016/09/28/144000/remedial-education/

Kim, J. E., Choi, Y., & Lee, C. H. (2019). Effects of climate change on Plasmodium vivax malaria transmission dynamics: A mathematical modeling approach. *Applied Mathematics and Computation*, *347*, 616–630.

Kung, D. (2023). How high school math can propel students to higher ed success. https://www.mathvalues.org/masterblog/how-high-school-math-can-propel-students-to-higher-ed-success

Larson, L. M., Pesch, K. M., Surapaneni, S., Bonitz, V. S., Wu, T. F., & Werbel, J. D. (2015). Predicting graduation: The role of mathematics/science self-efficacy. *Journal of Career Assessment*, *23*(3), 399–409.

Levy, R. (2015). 5 reasons to teach mathematical modeling. *Macroscope.* https://www.americanscientist.org/blog/macroscope/5-reasons-to-teach-mathematical-modeling

Liljedahl, P. (2021). *Building thinking classrooms in mathematics, Grades K–12.* Corwin.

Lim, K. H., & Selden, A. (2009, September). Mathematical habits of mind. In *Proceedings of the thirty-first annual Meeting of the North American Chapter of the International Group for the Psychology of Mathematics Education* (pp. 1576–1583). Georgia State University.

Martin, W. G., Carter, J., Forster, S., Howe, R., Kader, G., Kepner, H., Judith Reed Quander, J. R., McCallum, W., Robinson, E., Snipes, V., & Valdez, P. (2009). *Focus in high school mathematics: reasoning and sense making.* National Council of Teachers of Mathematics.

Mathematics Teacher: Mathematics Teaching and Learning PreK–12. (2023). Special Issue on Technology, *115*(5).

Matsuura, R., Sword, S., Piecham, M. B., Stevens, G., & Cuoco, A. (2013). Mathematical habits of mind for teaching: Using language in algebra classrooms. *Mathematics Enthusiast*, *10*(3), 735–776.

Matthews, T. D., & Banks, J. D. (2022). Culturally responsive teaching practices in the mathematics classroom. In T. M. Mealy & H. Bennett (Eds.), *Equity in the Classroom: Essays on Curricular and Pedagogical Approaches to Empowering All Students* (p. 48). McFarland & Company, Inc.

Moses, R. P., & Cobb,C. E., Jr. (2001). *Radical equations: Civil rights from Mississippi to the Algebra Project.* Beacon Press.

Nasir, N. S., & Hand, V. M. (2006, Winter). Exploring sociocultural perspectives on race, culture, and learning. *Review of Educational Research*, *76*(4), 449–475.

National Governors Association Center for Best Practices and Council of Chief State School Officers. (2010). *Common Core State Standards for Mathematics.*

National Research Council. (2001). *Adding it up: Helping children learn mathematics.* National Academies Press. https://doi:10.17226/9822

National Council of Teachers of Mathematics. (2000). *Principles and standards for school mathematics.* The Council.

National Council of Teachers of Mathematics. (2010). Essential Understanding series. The Council

National Council of Teachers of Mathematics. (2014). *Principles to actions: Ensuring mathematical success for all.* The Council.

National Council of Teachers of Mathematics. (2016). Moving students from remembering to thinking: The power of mathematical modeling. In C. R. Hirsch & A. R. McDuffie (Eds.), *Annual Perspectives in Mathematics Education, Mathematical Modeling and Modeling Mathematics* (pp. 100–101).

National Council of Teachers of Mathematics. (2018). *Catalyzing change in high school mathematics: Initiating critical conversations.* The Council.

National Council of Teachers of Mathematics. (2024). *Teaching data science in high school: Enhancing opportunities and success.* https://www.nctm.org/Standards-and-Positions/Position-Statements/Teaching-Data-Science-in-High-School_-Enhancing-Opportunities-and-Success/

NCSM: Leadership in Mathematics Education. (2019). *NSCM Essential actions: Instructional leadership in mathematics education.*

Ndaïrou, F., Area, I., Nieto, J. J., & Torres, D. F. (2020). Mathematical modeling of COVID-19 transmission dynamics with a case study of Wuhan. *Chaos, Solitons & Fractals*, 135, 109846.

Oakes, J. (1985). *Keeping track: How schools structure inequality.* Yale University Press.

Papert, S. (2006). *From the math wars to the new new math.* Keynote speech presented at the Seventeenth ICMI Study Conference: Mathematics Education and digital technologies, rethinking the terrain. Hanoi, Vietnam.

RAND Corporation. (2002). *Mathematical proficiency for all students: Toward a strategic research and development program in mathematics education.* https://www.rand.org/content/dam/rand/pubs/monograph_reports/MR1643/RAND_MR1643.pdf

Roschelle, J., Pea, R., Hoadley, C., Gordin, D., & Means, B. (2000, Fall Winter). Changing how and what children learn in school with computer-based technologies. *Future of Children, 10*(2), 76–101.

Sacristán, A. (2021). *Digital technologies, cultures, and mathematics education.* Proceedings of the Fourteenth International Congress on Mathematical Education. Shanghai, China.

Sacristán, A., Calder, N., Rojano, T., Santos-Trigo, M., Friedlander, A., & Meissner, H. (2010). The influence and shaping of digital technologies on the learning—and learning trajectories—of mathematics concepts. In C. Hoyles & J. Lagrange (Eds.), *Mathematics education and technology—rethinking the terrain* (17th ICMI Study) (pp. 179–226). Springer.

Sana, S. S. (2022). A structural mathematical model on two echelon supply chain system. Annals of Operations Research, *315*, 1997–2025. https://doi.org/10.1007/s10479-020-03895-z

Schielack, J., Charles, R., Clements, D., Duckett, P., Fennell, F., Lewandowski, S., Trevino, E., & Zbiek, R. M. (2006). *Curriculum focal points for prekindergarten through grade 8 mathematics: A quest for coherence.* National Council of Teachers of Mathematics.

Schmidt, W., Jorde, D., Cogan, L., Barrier, E., Gonzalo, I., Moser, U., Shimizu, K., Sawed, T., Valverde, G., Mcknight, C., Prawat, R., Wiley, D., Raizen, S., Britton, E., & Wolfe, R. (1996). *Characterizing pedagogical flow: An investigation of mathematics and science teaching in six countries*. Springer.

Schmidt, W. H., & Cogan, L. S. (1996). Development of the TIMSS context questionnaires. In M. O. Martin & D. L. Kelly (Eds.), *Third international mathematics and science study technical report: Vol. I. Design and development* (pp. 5-1–5-22). Boston College.

Schmidt, W. H., McKnight, C., Cogan, L. S., Jakwerth, P. M., & Houang, R. T. (1999). *Facing the consequences: Using TIMSS for a closer look at U.S. mathematics and science education*. Kluwer.

Schmidt, W., Mcknight, C., Raizen, S., Jakwerth, P., Valverde, G., Wolfe, R., Britton, E., Bianchi, L., & Houang, R. (2002). *A splintered vision: An investigation of U.S. science and mathematics education*. Springer Dordrecht.

Schoenfeld, A. H. (2020). Reframing teacher knowledge: A research and development agenda. *ZDM*, *52*(2), 359–376. https://doi.org/10.1007/s11858-019-01057-5

Sengupta-Irving, T. (2014). Affinity through mathematical activity: Cultivating democratic learning communities. *Journal of Urban Mathematics Education, 7*(2), 31–54. https://doi.org/10.21423/jume-v7i2a208

Shah, N. (2017). Race, ideology, and academic ability: A relational analysis of racial narratives in mathematics. *Teachers College Record*, *119*(7), 1–42. https://doi.org/10.1177/016146811711900705

Smith, M., Steele, M., & Raith, M. (2017). *Taking action: Implementing effective mathematics teaching practices Grades 6–8*. National Council of Teachers of Mathematics.

Stiff, L. V., & Johnson, J. L. (2011). Mathematical reasoning and sense making begins with the opportunity to learn. In M. E. Strutchens & J. R. Quander (Eds.), *Focus in high school mathematics: Fostering reasoning and sense making for all students* (pp. 85–100). National Council of Teachers of Mathematics.

Su, F. E., & Jackson, C. (2020). *Mathematics for human flourishing*. Yale University Press.

Sun, G. Q., Li, L., Li, J., Liu, C., Wu, Y. P., Gao, S., Wang, Z., & Feng, G. L. (2022). Impacts of climate change on vegetation pattern: Mathematical modeling and data analysis. *Physics of Life Reviews*, *43*, 239–270.

The University of Texas Charles A. Dana Center. (2019). *The Dana Center Mathematics Pathways*. https://www.utdanacenter.org/our-work/higher-education/dana-center-mathematics-pathways

U.S. Bureau of Labor Statistics. (2021). *62.7 percent of 2020 high school graduates enrolled in college, down from 66.2 percent in 2019*. https://www.bls.gov/opub/ted/2021/62-7-percent-of-2020-high-school-graduates-enrolled-in-college-down-from-66-2-percent-in-2019.htm

U.S. Department of Education. (2017). *Developmental education: Challenges and strategies for reform*. https://www2.ed.gov/about/offices/list/opepd/education-strategies.pdf

Wagner, D. (2019). Changing storylines in public perceptions of mathematics education. *Canadian Journal of Science, Mathematics and Technology Education*, *19*, 61–72. https://davewagner.ca/articles/2019_Wagner_CJSMTE_submitted.pdf

World Economic Forum. (2023). https://www.weforum.org/events/world-economic-forum-annual-meeting-2023/